WHOLLY
IRRESPONSIBLE
EXPLOITS!

D0785094

Sean Connolly will be familiar to listeners of BBC Radio Five Live and Radio Wales. Among his more than 50 books aimed at children and adults are *In Time of Need: Storms and Earthquakes* and *Witness to History: The Industrial Revolution*. He's also written for the *Kingfisher Science Encyclopedia*. His three children are either collaborators or guinea pigs, depending on the project.

WHOLLY IRRESPONSIBLE EXPLOITS!

65 WAYS TO MUCK ABOUT WITH SCIENCE

SEAN CONNOLLY

CORINTHIAN

Published in the UK in 2009 by Corinthian Books,
an imprint of Icon Books Ltd,
Omnibus Business Centre, 39–41 North Road,
London N7 9DP
email: info@iconbooks.co.uk
www.iconbooks.co.uk

Sold in the UK, Europe, South Africa and Asia
by Faber & Faber Ltd, Bloomsbury House,
74–77 Great Russell Street, London WC1N 3AU or their agents

Distributed in the UK, Europe, South Africa and Asia
by TBS Ltd, TBS Distribution Centre, Colchester Road
Frating Green, Colchester CO7 7DW

This edition published in Australia in 2009
by Allen & Unwin Pty Ltd,
PO Box 8500, 83 Alexander Street,
Crows Nest, NSW 2065

Distributed in Canada by
Penguin Books Canada,
90 Eglinton Avenue East, Suite 700,
Toronto, Ontario M4P 2YE

ISBN: 978-190685-001-2

Typesetting and design by Corinthian

Printed and bound in the UK by Clays Ltd, St Ives plc

ADULT SUPERVISION NEEDED

These experiments involve matches, hot liquids and ingredients that could be harmful if used incorrectly

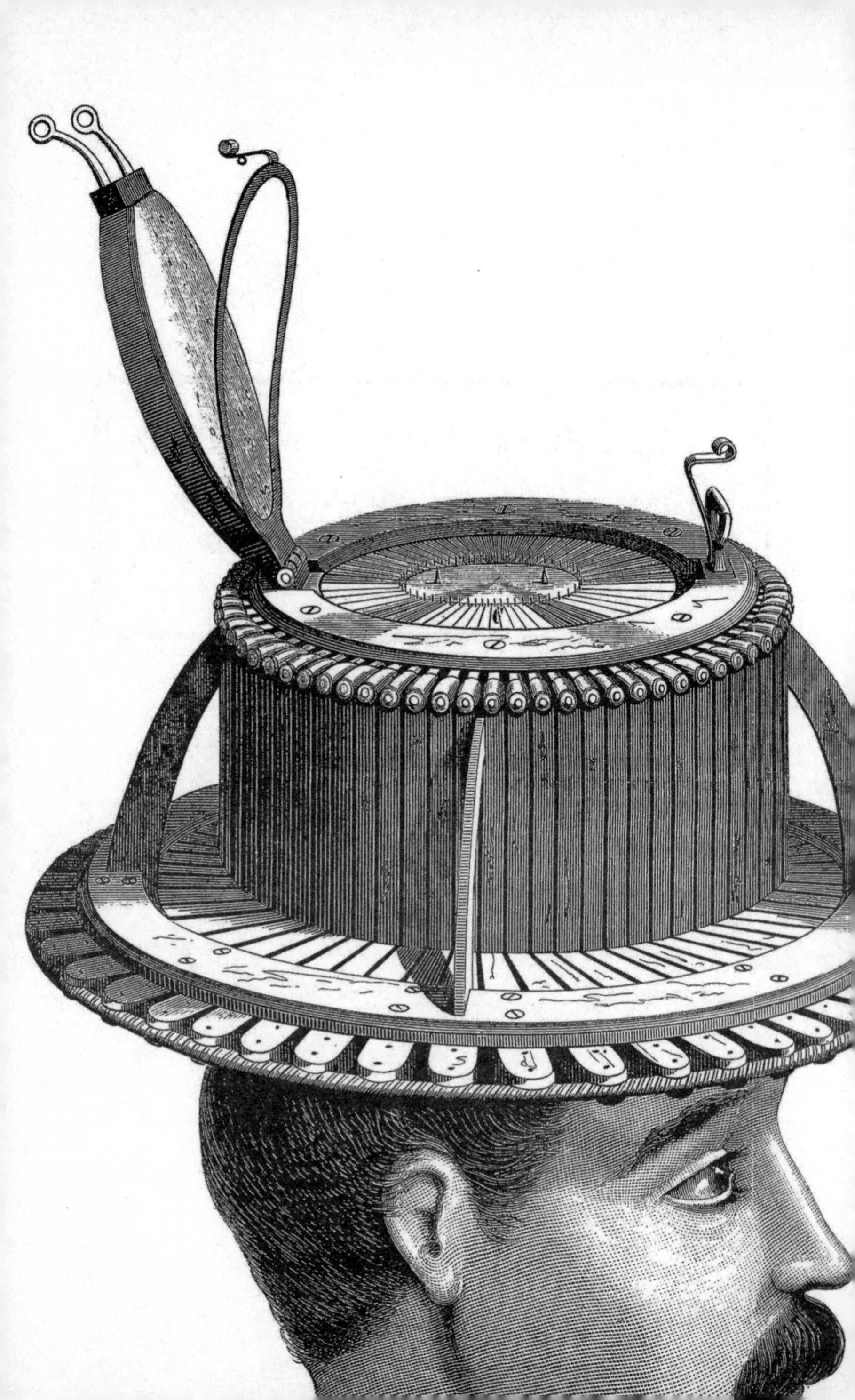

To my fellow travellers on this stage of a wonderful journey – Frederika, Jamie, Anna, Thomas and Dafydd

Contents

Introduction

The *Oxford Dictionary of English* defines science as 'the intellectual and practical activity encompassing the systematic study of the structure and behaviour of the physical and natural world through observation and experiment'. Its definition of exploit is 'a brilliant or daring achievement'.

So it is hardly surprising that people would be interested in something that combines intellectual study of the world around us with brilliance and daring. Just think of the Mongol warriors who swept through Asia and into Europe in the 13th century, brushing armies aside with their horsemanship and superb archery skills. They were defended by tight-fitting silk blouses that protected them from infection when they were hit by arrows.

Some wise Mongol must have deduced that silk would have this protective quality, enabling the rider to carry on until he reached safety. That's where intellectual and practical knowledge played a part. But it really became an exploit when the first warrior put this theory to the test, in a real battle. That took daring and brilliance. So did travelling in the first lifts in the 19th century and putting theories about powered flight into practice, as the Wright brothers did when they tried out the first aeroplane in 1903.

Wholly Irresponsible Exploits does not promise to help you conquer much of the known world, but it does tap into the traditions of curiosity and daring that have helped mankind through the ages. There's something thrilling about pushing boundaries until they stretch a little, and then justifying that push with an astounding result.

A noble tradition

Wholly Irresponsible Experiments took readers on a journey into this scientific wonderland, harnessing the ordinary and familiar to produce some marvellous payoffs. *Wholly Irresponsible Exploits* continues this journey, calling at some familiar stations and taking a few branch lines into uncharted territory. And like the first volume, it shows how the everyday world provides us with the tools to carry on with our scientific probing.

The 65 experiments described in the following pages use ingredients or materials that most households have, or which can be bought easily. And like the classic scientific experiments, which use questions as launch pads for enquiry, these experiments also seek to find and demonstrate answers. Some of the answers, however, might well tie in with a completely different set of questions. Along the lines of:

'Are you really pouring
light out of that thing?'

'Did you say you were going to fold an egg?'

'What on earth is holding those spoons up?'

'Wait a minute! Weren't you
just holding two beakers?'

In addition to the experiments, each of the eight chapters begins with a stirring account of a famous exploit in its field. A lot of these fall into the 'don't try this at home' category, which definitely does not apply to any of the 65 experiments in this book.

The 'i' word

All of this brings us to an important word in the title of this book – 'Irresponsible'. Where does being irresponsible tie in with conducting experiments? Surely it's the opposite of the scientific method? In response we can cite former US President Bill Clinton and his famous phrase: 'That depends upon what your definition of "is" is.' Well, if 'is' can carry with it more than one meaning, then it's hardly surprising that there could be more than one reading of the word 'irresponsible'.

Although using the 'i' word, *Wholly Irresponsible Exploits* advocates due care and attention in each experiment. The presentation of each experiment is straightforward and logical, right down to any words of special warning (in the Take care! section) that apply to the experiment. Instead it calls on readers to throw off some of the shackles of being a grown-up and find the child within us all.

No child, after all, would worry about flinging eggs at a clean bed sheet. And that inner child would hardly prefer sorting out an income tax return (or any other 'responsible' activity) to letting nature do its bit to snap two school rulers. So, by that definition, each of the experiments certainly does merit the descriptive term 'irresponsible'.

Who can do these experiments?

Any book that appeals to the 'child within us' is bound to appeal to children themselves, and this is no exception. We heartily recommend demonstrating each experiment to budding young scientists, and some of the experiments can involve children as volunteers or even participants. But bear in mind that the responsibility for each experiment lies with the adult conducting it. These experiments are *for* children as well as adults, but they are not to be conducted *by* them.

The final section of each experiment, Take care!, highlights any particular warnings relevant to the experiment. Some of these are no more than bits of friendly advice on how to get the best effects. Others have a more practical aim of drawing the reader's attention to ingredients or actions that call for extra care. A special Match alert is a prominent flag to any experiment that involves matches or an open flame.

How this book works

The 65 entries in *Wholly Irresponsible Exploits* are grouped in seven chapters, each representing a different scientific theme or intended result.

A typical entry introduces the nature of the experiment and what to expect, before breaking it down into the following sections.

YOU WILL NEED – A straightforward list of ingredients.

METHOD – Numbered step-by-step and easy-to-follow instructions.

THE SCIENTIFIC EXCUSE – The *raison d'être* for the experiment – or possibly your hurried explanation to an impatient or angry spouse!

TAKE CARE! – Special advice (and in some cases, warnings) for the experiment.

At a glance

The 65 experiments have also been grouped at the back on page 187 according to how long it takes to complete them – from the first stage of preparation to the 'oohs' and 'aahs' at the conclusion. You might have a whole Saturday at your disposal or only a few minutes free.

The categories here help you to choose an experiment perfect for the time you have spare.

Flash in the Pan – less than 2 minutes

Five-minute wonders – 2–5 minutes

On the hour – up to 1 hour

The 8-hour day – 1–8 hours

Going the distance – a full day or more

Isn't it time you went out and lit some weed (waterweed, of course)? Or maybe you're worried that your grapes don't glow enough? Perhaps you've decided you're ready to boil some water, *in a paper cup*! The following pages will let you do all of these things, and much more, all in a spirit of playful scientific enquiry.

For most of the experiments, a broad smile and an open mind will count for far more than a white coat and a calculator. So throw yourself into these funny, eye-opening, quirky experiments and see where they take you. And in the process you'll have a chance to learn – and maybe even teach others – a little science!

That's Life!

In an age when green issues are on everyone's mind, it's only natural to turn to nature for inspiration. Whether it's in the form of a grow-your-own experiment or harnessing the power of your favourite fruit, the natural world is the place to see scientific work in progress. Scientists have taken centuries to understand and recreate some of the marvels that occur naturally in the world around us. But be careful – there can be a real sting in the tail with some of this research.

THE BEE BEARD

It takes a special type of courage – or foolhardiness – to work with some animals. People seem unfazed by the idea of 'horse whisperers', seal trainers or even lion tamers. But dealing with some other species seems to be risking big trouble. An obvious example of a 'you wouldn't catch me doing that' species is the humble bee. Most people shudder and squirm at the idea of dealing with these intelligent, co-operative insects.

Despite that perception, or maybe even because of it, some people continue to surround themselves with a honey-making workforce. Most of us have seen bee-keepers in their baggy spacesuit-style outfits. Even with these head-to-toe coverings, though, beekeepers seem to be courting trouble. It's harder still to imagine stripping off almost everything and getting 'stuck in' with a swarm of buzzing, crawling bees.

That is exactly what people do when they set about creating a bee beard – attract the largest possible number of bees to land and remain on their body. The bee beard is one of the acknowledged Guinness World Record categories. In June 2005, Irish bee-keeper Philip McCabe set out to break that record.

McCabe stripped down to a pair of briefs. He put on goggles, wore a mouth mask and put tissues in his nostrils for protection. To attract bees in their thousands, he hung a cage containing a queen bee round his neck. Assistants covered his naked skin with honey and then 'poured' bees from nearby hives through a funnel onto McCabe's chest. Thousands of the bees began to crawl up his body and attach themselves to his chin and neck, creating the beard. Eventually, his whole upper body was covered.

It was certainly dramatic, but was it a record? Luckily, no one really needed to do the count, bee by bee. Instead, record-keepers estimated the number by weight. The world record of 350,000 bees weighed 40 kg; this figure was reached by weighing the person before and after the bees had swarmed over him. After two hours of standing, McCabe had become numb and called for a weigh-in. Despite the drama, McCabe could only attract 200,000 bees. He blamed the shortfall on a brisk Irish breeze, which kept blowing the bees off his torso.

The Flame Tree?

We all have an idea that some sort of chemical reaction takes place to help plants produce their own food. Now do they use oxygen to produce carbon dioxide, or is it the other way round? This experiment will help you finally see the light.

YOU WILL NEED

- Large beaker (about 1 litre capacity)
- Test tube
- Glass funnel (with head small enough to fit inside the beaker)
- Long 'fire-starter' matches
- Waterweed (available at aquarium-supply shops)
- Water
- A friend to help with the 'payoff' of the experiment

METHOD

1. Put enough waterweed into the beaker to cover the bottom.
2. Fit the funnel, wide side down, over the waterweed.
3. Make sure that some strands of waterweed jut out from inside the funnel, creating a small gap above the base of the beaker.
4. Fill the beaker with water up to 2 cm from the top of the funnel.
5. Fill the test tube with water and cap your thumb over the end to stop spills.

6. Quickly but accurately fit the test tube down over the upturned funnel.
7. Place the set-up carefully in direct sunlight and let it stand for an hour; observe it but leave it alone. You should see bubbles rising inside the test tube.
8. After about three hours the bubbles should have driven all the water from the test tube.
9. Carefully remove the empty test tube and hold it upside down.
10. Ask your friend to light a fire-starter match, then blow it out and then insert the burnt end inside the test tube. The match should re-ignite inside the tube.

— THE SCIENTIFIC EXCUSE —

This experiment is a fine example of photosynthesis, the process by which plants use light and water to produce the carbon dioxide they use as food. Oxygen, a by-product of this process, is released and bubbles up into the test tube. It reacts with the wood to make it combust (burst back into flame).

TAKE CARE!

It is important to keep air from the water-filled test tube, so you might want to practise your technique before performing the experiment. And given that the experiment depends on photosynthesis for its effect, try to do it on a sunny day.

MATCH ALERT!

This experiment involves the use of matches and should only be conducted by a responsible adult.

GOING GREEN – OR IS THE GREEN GOING?

Each autumn, hundreds of thousands of visitors from around the world travel to the American region of New England to see the 'fall colours'. The leaves on the trees in New England forests lose their green colouring to reveal some splendid hues. Here's a chance to mimic Phase One of that process, and maybe a little more. Is that a coach tour drawing up out in front?

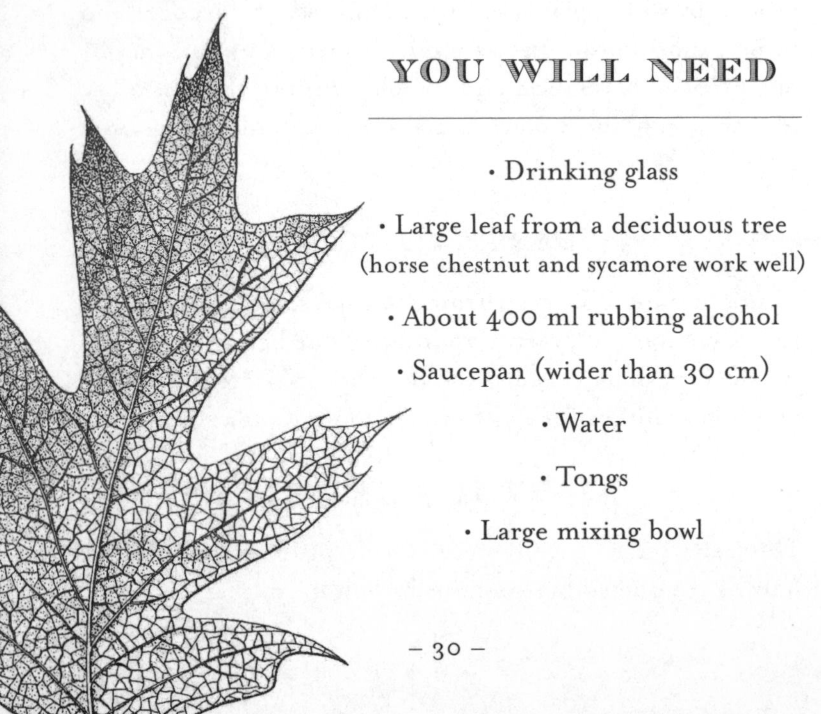

YOU WILL NEED

- Drinking glass
- Large leaf from a deciduous tree (horse chestnut and sycamore work well)
- About 400 ml rubbing alcohol
- Saucepan (wider than 30 cm)
- Water
- Tongs
- Large mixing bowl

METHOD

1. Half fill the saucepan with water and bring to the boil.
2. Add the leaf to the saucepan and simmer gently for two minutes.
3. Turn off the heat, remove pan and use the tongs to transfer the leaf to the glass. Keep water in pan.
4. Fill the glass (now containing the leaf) with rubbing alcohol.
5. Add cold water to the saucepan until it has cooled enough to be touched.
6. Place the glass in the mixing bowl and add warm water from the saucepan to the bowl until it is almost as high as the top of the glass. Leave for one hour.
7. When you return, the alcohol will have turned green.
8. Use tongs to extract the leaf: it should have lost most of its green colouring.

THE SCIENTIFIC EXCUSE

Chlorophyll is the pigment that helps plants use light to produce their own food. The word comes from the Greek words *chloros* ('green') and *phyllon* ('leaf'). The rubbing alcohol, working even faster because it is surrounded by warmth, breaks down some of the cell walls of the leaf. The results are a greenish liquid in the glass, thanks to the chlorophyll, and the 'de-greened' appearance of the leaf.

TAKE CARE!

The two liquids in this experiment – boiling water and rubbing alcohol – come with an 'adults only' warning. Children should only be witnesses here.

PINEAPPLE POWER

Have you ever wondered why your tongue feels a bit rough and raw after you've eaten fresh pineapple? Sometimes it feels as though it is the pineapple that is doing the eating – consuming the outer layer of your tongue. In fact, that's a pretty accurate description of what happens, and this experiment shows how.

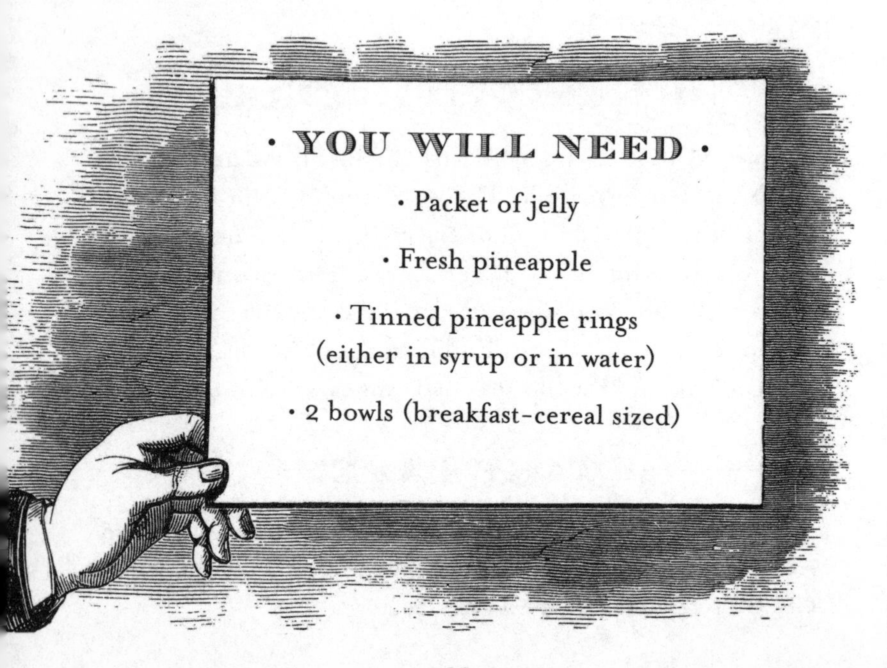

· YOU WILL NEED ·

- Packet of jelly
- Fresh pineapple
- Tinned pineapple rings (either in syrup or in water)
- 2 bowls (breakfast-cereal sized)

METHOD

1. Prepare the packet of jelly according to its instructions and fill the two bowls before the mixture begins to set.

2. Allow both bowls of jelly to set.

3. Cut a ring from the fresh pineapple and remove a ring from the tinned pineapple.

4. Put the fresh ring on top of one bowl of jelly and the tinned ring on the other.

5. Wait about 15 minutes and observe.

6. By that time the fresh pineapple ring will have eaten its way down through the jelly but the tinned ring will remain on the surface.

THE SCIENTIFIC EXCUSE

This appetising experiment demonstrates the effectiveness of enzymes, proteins that catalyse (speed up) chemical reactions. Fresh pineapple contains the enzyme bromelase, which digests the gelatine (which is also a protein). That funny feeling on your tongue is also the result of bromelase working away. Tinned products such as pineapple are heat-treated, which kills off the enzymes such as bromelase. That is why the ring of tinned pineapple sits obediently on the top of the jelly.

TAKE CARE!

Take care cutting the ring from the fresh pineapple: it is better to have an adult carve the ring and remove the outer skin with a sharp knife.

THE PEANUT HARVEST

Maybe doctors would be less worried about children becoming fat if we all had to wait months to prepare our snacks. So in that respect, this exploit could almost be called responsible. Or maybe not, if you have to tell your Christmas guests that they'll have to wait another week or two for their dry roasted peanuts to ripen.

· YOU WILL NEED ·

- Fresh peanuts (available from health food shops)
- Sandwich bag or small glass jar with lid
- Paper towel
- Water
- Soil
- Plant tray (30 cm x 45 cm)

METHOD

1. Carefully remove the shells from one or two peanuts, making sure to keep the red skin on the nuts inside.

2. Dampen three sheets of paper towel and add these to several nuts.

3. Put this mixture carefully inside the sandwich bag or jar, trying to keep parts of the nuts in view.

4. Seal the sandwich bag or screw on the lid of the jar.

5. Watch for about a week, and then you should see the nuts sprouting.

6. Fill the plant tray with soil and make a hole large enough (about 3 cm deep) to plant the flowering nut.

7. Keep the pot watered regularly as the peanut grows into a bush, which will be about 45 cm tall. After seven or eight weeks, it will have many yellow blossoms.

8. The blossoms will fall off and the stalks (with pods at the end where the blossoms were) droop towards the ground and then tunnel beneath the soil.

9. After another nine weeks you can dig up the plant – peanuts will have matured underground where the pods were.

— THE SCIENTIFIC EXCUSE —

This absorbing (and ultimately satisfying) project demonstrates something that most people don't realise – that peanuts are not nuts at all. Nuts are the dry seeds of plants and never grow underground. Peanuts, strictly speaking, are in another plant category called legumes and are related more to soya beans and peas.

TAKE CARE!

Make absolutely sure that no one taking part in this experiment – or eating its results – has a peanut allergy.

Water Great Idea

Water, water everywhere,
And all the boards did shrink;
Water, water everywhere,
Nor any drop to drink.

Could the great poet Samuel Taylor Coleridge have been writing about the following chapter, which uses water as the source of scientific magic? Maybe not, but think of how useful he'd have found a DIY ocean – and a mum to row across it with him.

Dan's Row with His Mum

The idea of crossing the Atlantic – especially from east to west – has captured people's imaginations for centuries. Debate rages about who 'discovered America'. Was it Columbus in 1492? Leif Erikson and his Viking voyagers 500 years earlier? Or maybe Saint Brendan from Ireland and his band of monks, sailing in small ox-hide vessels in the sixth century?

Just as tantalising are the stories of Atlantic crossings after Columbus – the fastest, the first single-handed, the youngest single-handed, the first in a rowing boat. Most of these chronicle solitary exploits, or shared efforts by tough sailors. None of them mentions a mariner rowing across the Atlantic with *his mother*.

That is exactly what Dan Byles did in 1997. Byles, 23 years old at the time, had learned of the first Atlantic Rowing Race, founded by British sailing hero Chay Blyth, in 1995. At the time, more people had climbed Everest and been in space than had rowed across the Atlantic: six had died trying. Byles and a friend began training, but work commitments soon ended the friend's involvement.

Dan did what young men often do in times of need – he turned to his mother for help. Jan Meeks, 50 and recently widowed, was looking for ways to keep her spirits up. Her son's offer was unexpected – especially as Jan preferred quiet evenings in, listening to classical music. Her only rowing experience had been on the Serpentine when she was eight. But she agreed, and the pair spent 18 months training for the crossing. Jan had to remortgage her house to build the 6.5-metre wooden boat.

The race was as challenging as they had expected. Equipped with a £30 camping stove, no satellite phone and only one set of foul-weather gear between them, they set off on the 4,900-km voyage from Tenerife to Barbados. They encountered whales, sharks and waves as large as a house. Jan joked that the boils she developed made her bottom look like a pizza.

After 101 days at sea, they crossed the finishing line. They came 22nd out of the 22 teams to complete the race, but they had set two records: as the first mother-and-son team to row across the Atlantic, and as the oldest person to row across the Atlantic.

BACK-UP PLAN

'Darling, the toilet is backed up.' There's a phrase designed to strike terror into the heart of the average newspaper-reading husband. Somehow, 'darling, the glass has backed up' lacks a bit of edge. But what it lacks in shock, it makes up for in wonder. Here is an experiment that deserves an audience: even Dad might come willingly when he learns he's only a bystander.

YOU WILL NEED

- Shallow baking tray or aluminium pie plate
- Candle
- Matches or lighter
- Water
- Drinking glass or flask (about 350 ml capacity)

METHOD

1. Light the candle and drip some wax onto the centre of the baking tray or pie plate.
2. Anchor the still-burning candle in the wax before it hardens.

3. Fill the tray or plate about halfway up with water.

4. Place the glass or flask upside down over the candle.

5. After a few seconds, the candle will burn out, then water will rise up into the upturned glass and remain at a higher level than in the surrounding pan.

THE SCIENTIFIC EXCUSE

Air pressure – or more precisely, different air pressures – do the work of this experiment. By placing the cup over the burning flame, you help the air inside warm up. But it is a limited supply of air, and the candle soon burns up all the oxygen from the air ... and in short order the air begins to cool. That is where the pressure comes in: cooler air exerts less pressure than the previously-warmed air. The water rushes in because there is less air pressure to hold it back.

TAKE CARE!

Make sure you drip enough wax to support the candle securely, since a sudden bump could cause an unstable candle to wobble and tip over. Try adding a drop or two of food colouring to the water in case people need to see the experiment from a little distance.

MATCH ALERT!

This experiment involves the use of matches and should only be conducted by a responsible adult.

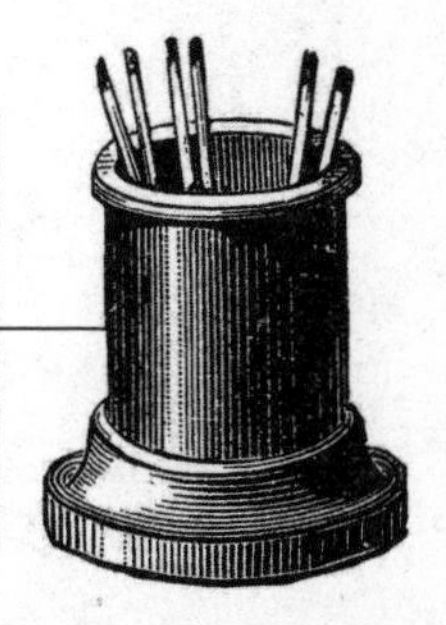

The Magic Circle

If only every household object acted as obediently as the little piece of string in this experiment. Life would be so much easier if we could wave a cotton bud at some batter and it immediately became a cake, or if we could touch a sheet and the bed made itself. These breakthroughs will come in good time, no doubt. In the meantime, try to make do with this little bit of magic.

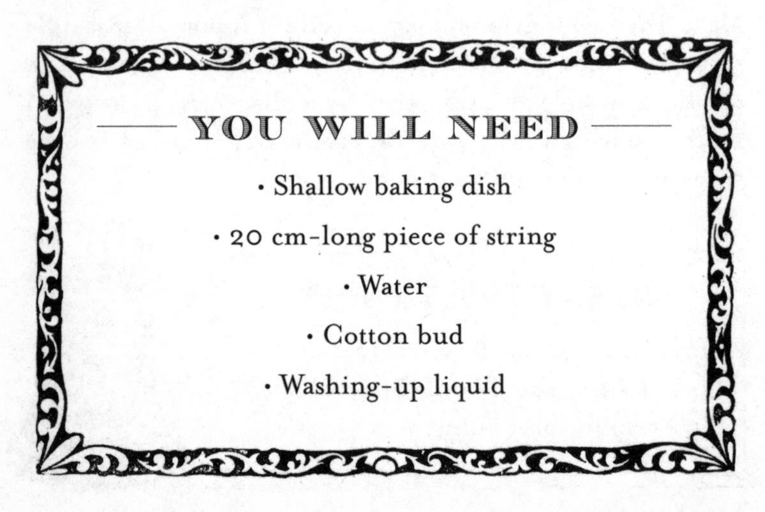

You Will Need

- Shallow baking dish
- 20 cm-long piece of string
- Water
- Cotton bud
- Washing-up liquid

METHOD

1. Fill the baking dish about halfway up with water.
2. Tie the ends of the string to make a circle.
3. Float the string on the water; it will lose its circular shape.
4. Dip a cotton bud in some washing-up liquid and then dip this soapy end inside the floating string.
5. The string will quickly become a perfect circle.

THE SCIENTIFIC EXCUSE

We often read about something 'breaking the tension', and in this case it is the washing-up liquid that does just that. The molecules along the top of the water are held together – making a sort of 'skin' – by surface tension. The soap, however, is able to pierce between the water molecules and break this skin. The newly freed liquid particles move out from the soap. Because they exert a similar force along the entire inside of the string, they force it into an exact circle.

TAKE CARE!

There's no real need to worry with this experiment. Just make sure no one mistakes the soapy solution for water when you've finished.

CORK ON A BENDER

You might be feeling a little the worse for wear yourself, if you overplay your hand with this display and come across as too much of a wizard. At its heart, of course, is a sound scientific principle, but without test tubes and Bunsen burners to act as props, your audience might think some other force is at work here. Bear in mind that this is one of the simplest and quickest experiments you can possibly perform, so enjoy it.

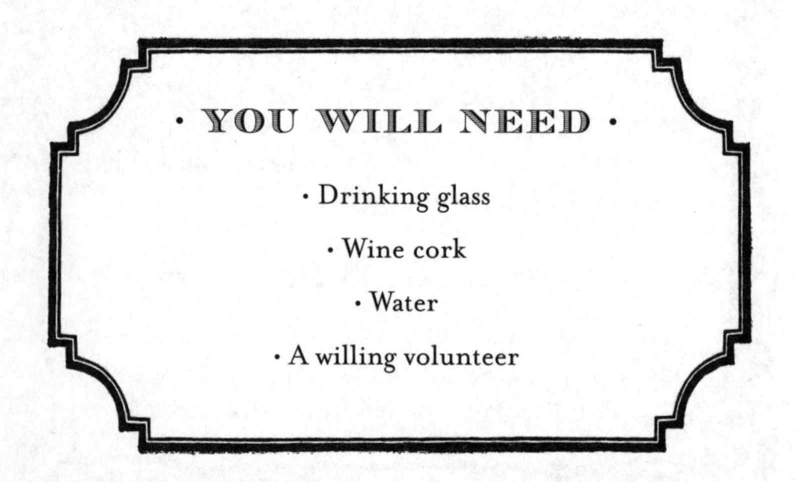

· YOU WILL NEED ·

- Drinking glass
- Wine cork
- Water
- A willing volunteer

METHOD

1. Fill the glass almost to the top with water.

2. Ask a volunteer to place the cork so that it floats in the centre of the water. (It won't, and will keep drifting to the side.)

3. Explain that you will succeed where they have failed; take the cork out and fill the glass right to the top with water.

4. Now, you take your turn and show how the cork can remain exactly in the centre as you had planned.

THE SCIENTIFIC EXCUSE

When the glass was partially filled, the water clung slightly to the inside of the glass, causing a slight lip to form around the edge of the water. The cork, being buoyant, simply floated to the highest point – the edge. The 'full' glass, on the other hand, is slightly fuller than full because water tension causes the surface to bulge. Now the highest point is in the centre, at the top of the bulge. And that is where the cork will remain.

TAKE CARE!

The biggest risk you'll face is from your frustrated volunteer, who might want to exact some sort of revenge. Watch your back!

– The – NO–FLOW SIEVE

Everyone has a good idea of how a sieve works, whether it's used for panning for gold or preparing pasta. It's simple: water or some other liquid flows through freely, leaving solids (which don't fit through the gaps) behind. But what about a sieve that won't let any water through, even when you know there are some gaps in there? Is there a gap in your knowledge, or is there something more to it?

• YOU WILL NEED •

- 20 cm x 20 cm piece of wire gauze (as used for mosquito-proofing)
- 1-litre glass bottle or jar (ideally with mouth about 6–8 cm wide)
- Strong elastic band
- Water
- A friendly volunteer

METHOD

1. Hold the piece of wire gauze up and blow through it (to silence doubters later).

2. Fill the bottle right to the top with water.

3. Fit the gauze over the mouth of the bottle and secure it with the elastic band.

4. Ask your friend to put his hand over the top of the bottle and to turn it upside down, still holding his hand over the mouth.

5. Now ask your friend to position the bottle, still upside down, over your head and then to pull his hand away quickly.

6. No water will have poured out of the bottle, and your head should stay dry.

THE SCIENTIFIC EXCUSE

Surface tension, the force that holds molecules together, helps water form a 'skin' when it comes into contact with air. This skin anchors itself on solid objects such as the side of a glass, or in this case the tiny metal cross-bars that make up the gauze. Hundreds of these tiny water barriers (held firm with surface tension) keep the water from flowing out when the bottle is turned upside down.

TAKE CARE!

The directions above lead to a low-risk result – even if the experiment failed, the only person to suffer would be you. On the other hand, the same display becomes really irresponsible if you try it on someone else – a grumpy uncle, perhaps?

SILENT ERUPTION

Every now and then, you will come across an experiment that will have you shaking your head and wondering, 'Did I really make that happen?' Usually this kind of head-scratching demonstration is the sort that comes with very little equipment and even less in the way of scientific props. Sit back now, and watch this pair of very ordinary bottles – with even more ordinary ingredients – perform an extraordinary feat.

YOU WILL NEED:

- 2 identical 1-litre glass bottles (each with mouth about 6–8 cm wide)
- Piece of card about 10 cm square
- Hot water
- Cold water
- Ink

METHOD

1. Fill one bottle with cold water and the other with hot water (both from the tap).

2. Add some ink to the bottle of hot water and swish the bottle round so that the colour is uniform.

3. Make sure that each bottle is completely full; top up if necessary.

4. Press the piece of card to the top of the cold-water bottle.

5. Keeping the card firmly in place, turn the 'cold' bottle upside down so that it rests exactly above the top of the 'hot' bottle. (The card is lodged between them.)

6. Holding the bottles firmly, slide the card out from between them.

7. The ink-coloured hot water will rise into the top bottle and clear cold water will flow down into the other bottle.

THE SCIENTIFIC EXCUSE

The key to this experiment is buoyancy, the upward force that keeps ships – and swimmers – afloat. Warm water rises within a closed system just as warm air rises outside. In this case, the warmer water is easy to identify because of the ink, but the clear, cooler water is just as recognisable as it eases past the inky water to rest at the bottom of the lower bottle.

TAKE CARE!

Mastering the 'bottle-flip' can take time. It's worth having a few test runs, first with empty bottles and then with cold water in both, until you feel you can do it easily.

– The – Floating Needle

Here's a quick test. Put a piece of tissue and a sewing needle in a glass and one of them will float. But which one? Try this experiment to get a real eye-opener about water power. You might want to make a friendly bet with your audience as well!

YOU WILL NEED

- Drinking glass
- Piece of tissue
- Small sewing needle
- Water

METHOD

1. Fill the glass right to the brim with water.

2. Tear a square of tissue just smaller than the size of the glass.

3. Carefully lay the needle on the tissue and lower them both down to the surface of the water.

4. The tissue will float, with the needle on it, before sinking down …

5. … but the needle will remain floating on the surface of the water.

THE SCIENTIFIC EXCUSE

Here's another demonstration of surface tension, and maybe one that will stick in people's memories because it seems so wrong – until they think it through. Most of a tissue is actually air, and once the air is driven out of the tissue (as it gets soaked) most of what is left is water. The sodden tissue sinks to the bottom of the glass but the water molecules by then have regrouped, re-establishing surface tension which keeps the needle floating.

TAKE CARE!

Be careful handling the needle and make sure it is put away when the experiment is finished.

GHOSTLY SKATER

For thousands of years, scientists and philosophers have tried to find the secret of perpetual motion. That mysterious goal has remained elusive, despite mankind's progress in science and technology. But maybe they've all been looking in the wrong places – in distant galaxies or in the nano-worlds of electron microscopes. What if the secret were closer to hand ... maybe on the kitchen counter?

YOU WILL NEED

- 1 cm-thick slab of cork (about 12 cm square)
- 4 double-ended needles (about 6 cm long)
- 3–4 camphor mothballs
- 30 cm-wide bowl
- Card or construction paper
- Coloured pencils
- Knife
- Water
- Rubber gloves

METHOD

1. Draw a figure of a twirling skater on the card; cut out and colour in this shape, leaving a small tab extending from the bottom.

2. Cut a 5–6 cm-wide circle and four 2 cm squares from the cork.

3. Make a small slit in the centre of the cork circle and slip

the tab (from the base of the skater) into it. The skater should be upright.

4. Cut 5 mm-wide V-shapes from one side of each of the cork squares. Put on the rubber gloves for the remaining steps.

5. Cut a small piece of mothball to fit in each of the V-shapes.

6. Fit a needle into each cork square, on the opposite side to the mothball.

7. Fit the other end of each needle into the edge of the cork circle, so that they form a cross shape with the cork in the centre.

8. Fill the bowl almost to the top and place the 'cork cross' on it.

9. The skater will twirl and twirl.

— THE SCIENTIFIC EXCUSE —

This experiment is a bit of a change from some of the others in this chapter because it works by *reducing* the surface tension of the water. The camphor in the pieces of mothball slowly dissolve and create a camphor solution as they mix with the water. This camphor solution has less surface tension than normal water. So, the 'pull' of the water on the leading side of each cork rectangle is stronger than the pull from the mothball side. Put simply, the stronger pull wins and the corks go round and round ... as does the spooky skater above.

TAKE CARE!

Wearing rubber gloves protects your hands from the camphor, which can irritate the skin. Take great care in cutting out all the shapes; if a sharp knife is used for this cutting, make sure an adult does it.

The DIY Ocean

Pity the poor children who spend their childhood well away from the sea, never hearing the rhythmic lap of waves against the shore. This quick and easy experiment gives them a glimpse of the seashore, minus the sunburn and sand flies. As for rhythmic lapping, you might still have to rely on the family dog.

• YOU WILL NEED •

- Empty 2-litre plastic bottle (with cap)
- Funnel
- Blue food colouring
- Water
- Cooking oil (sunflower oil works well)

METHOD

1. Fill the bottle about a third of the way up with water.

2. Add several drops of food colouring and swirl the bottle a few times to mix it in.

3. Use the funnel to fill the rest of the bottle with cooking oil.

4. Cap the bottle tightly.

5. Hold the bottle horizontally, raising and lowering one end.

6. The dyed water will seem to make blue waves as it rocks back and forth.

THE SCIENTIFIC EXCUSE

This mesmerising display of wave power is a wonderful demonstration of how oil and water do not mix. Bear in mind two important points. First, a denser liquid (in this case water) will sink below a lighter one. But that doesn't explain why the two liquids don't just mix together to form a new one, like white coffee. That is because the oil is not strong enough to break the powerful hydrogen bonds that hold the water molecules together. The two liquids are destined to remain forever separate.

TAKE CARE!

Make sure the cap remains tightly sealed throughout this experiment, as otherwise the clean-up could be lengthy.

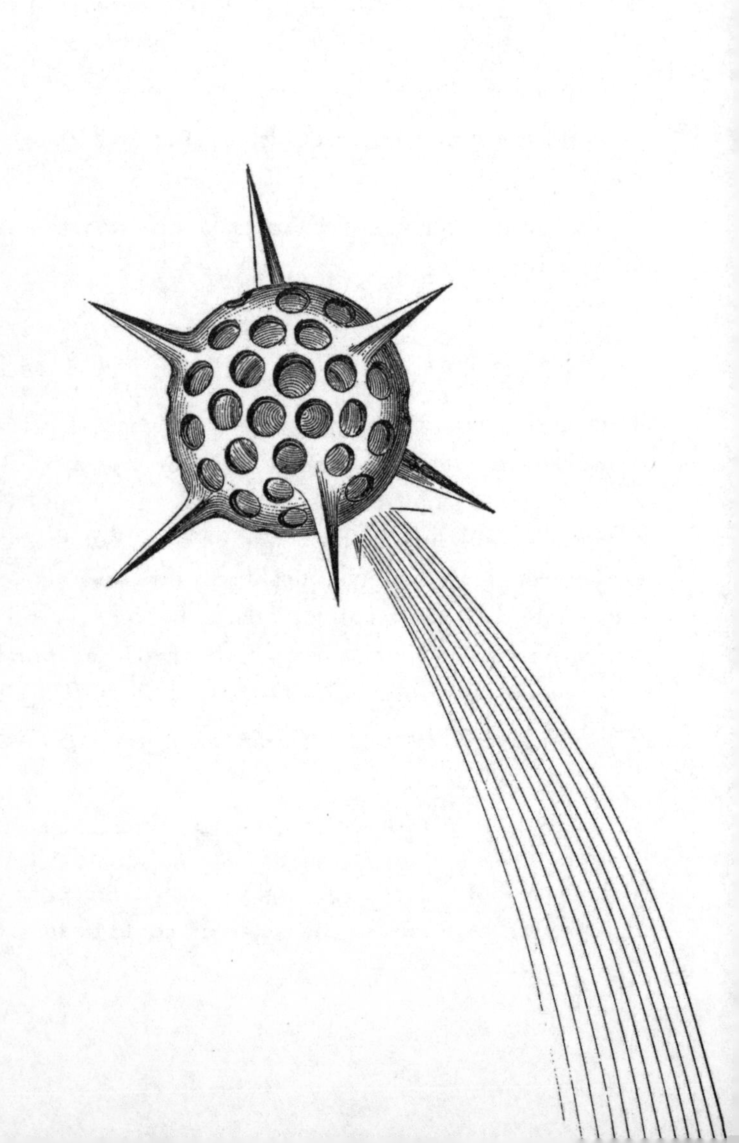

Sparks Will Fly

Scientists tend to attract more people with their magnetic research than their magnetic personalities. Or do they? Who says you can't manage both? After all, it seemed to work for wise old Ben Franklin, who managed to guide America to independence and to charm the French court, when he wasn't busy discovering the secrets of electricity. The following experiments will certainly generate a lot of interest, provided you stay positive.

THE KEY DATE FOR ELECTRICITY

What is it about wild weather that brings out the daredevil in some of the wisest people? Scores of ordinary nine-to-five types across America spend every spring poised to abandon their work cubicles at the drop of a hat. Then they jump into their cars to become 'storm chasers', tracking severe thunderstorms that they hope will develop into tornadoes.

More than 250 years ago, one of America's wisest statesmen risked electrocution in a similarly rash exercise – his aim was to unlock the secret of lightning. Benjamin Franklin did not have an established track record in death-defying stunts. He was more famous for his wise sayings (in *Poor Richard's Almanac*), the invention of bifocals and his diplomatic efforts in helping the United States become independent.

By the 1750s Franklin, like other scientists,

had become convinced that lightning was really the same as static electricity, but millions of times more powerful than the shocks that people get from woollen jumpers. He also knew that lightning strikes – or seeks out – high objects. The third thing that Franklin must have known – because he could not have survived his famous experiment in the face of a full-on electrical storm – was that the static electricity charge must build up as the storm brews.

With these things in mind, Franklin and his 21-year-old son William noticed dark clouds developing over Philadelphia one June afternoon in 1752. They launched a special kite that had a metal key at the end of its string. Extending from the suspended key were a ribbon (which Franklin held) and a wire leading down to a Leyden jar, which stores electrical charge.

Benjamin and William waited patiently as several clouds passed nearby: nothing happened to the kite or the key. Then Benjamin noticed that the strands of the string leading down from the kite had become wet, and possibly charged (an electrical charge would travel along a wet string far more easily). He held out his free hand to the key and as it drew near, there was a bright flash. It was instant proof that the kite had drawn some of the electrical charge downwards.

The Paper Polka

Here's another one to file under 'why didn't I think of that?' We've all learned about positives and negatives, attraction and forces, but at times it feels a little bit too confined to the textbook. Is there any way to study this sort of thing that's a bit livelier, maybe lively enough to set your toes tapping? Well, maybe there is.

YOU WILL NEED

- Plastic box with clear, snap-shut lid (the ideal box would be shallow and the size of a shoebox)
- Paper confetti (or cut-up bits of paper from a shredder)
- Piece of cloth or wool

METHOD

1. Pour the confetti or cut-up paper into the box so that the bottom is not quite covered.

2. Snap the lid shut.

3. Rub the cloth or wool along the top of the lid.

4. The bits of paper will dance up and down inside the box.

THE SCIENTIFIC EXCUSE

This experiment is all about electrical charges and the properties of attraction and repulsion. Rubbing the surface of the lid imparts an electrical charge, which attracts the bits of paper. They rise up to meet the lid and then pick up the charge from it. In an instant it's a case of like pushing away like (the paper and lid having the same charge), so the paper is forced back. When they hit the bottom of the box, they lose that charge and are attracted back up to the lid.

TAKE CARE!

The modest scale of the electrical charge means that this experiment is harmless.

Spooky Calling Card

As Professor Liedenbrock and his anxious nephew Axel are beginning to feel lost among the endless underground tunnels in Jules Verne's *Journey to the Centre of the Earth*, their torch briefly illuminates two letters, 'AS', on the cavern wall. This sight brings them enormous relief, because they know they are indeed following in the footsteps of the great Danish explorer Arne Saknussemm. The image of the letters forming seemingly out of nowhere is startling, and enough to set any scientist wondering whether they too could pull off a clever trick with letters. Here's one way to do it.

You Will Need

- Plastic lid from a margarine or spreadable butter tub
- Card
- Sticky tape
- Scissors
- Pepper
- Pencil
- Some flannel

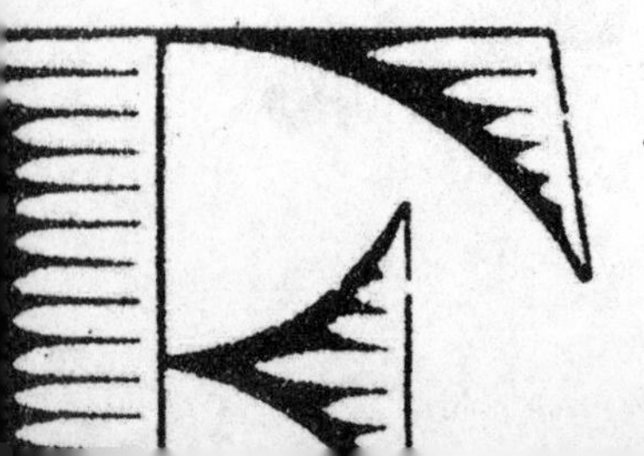
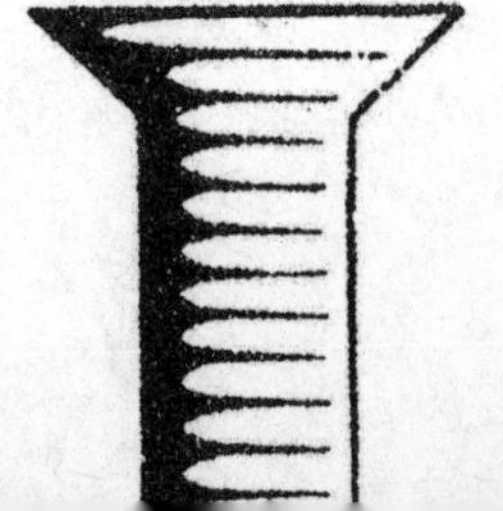

METHOD

1. Mark out a letter, about half the size of the margarine lid, on the card. This will become a stencil, so use a letter such as 'E', 'L' or 'V' rather than one that has a hole in it (such as 'O' or 'P').

2. Ensure that the curve or the lines forming this letter are 1 cm wide, and then cut out the shape.

3. Tape this stencil to the underside of the plastic lid.

4. Rub the stencil letter (i.e. the bits of plastic showing through the stencil) with the flannel for about ten seconds.

5. Taking care not to touch the bit you just rubbed, peel off the stencil and remove it.

6. Shake pepper over all the underside of the lid so that it forms a thin layer. Shake the lid to distribute it.

7. Flip the lid over so that the excess pepper falls off.

8. Flip the lid back to reveal the spooky letter.

THE SCIENTIFIC EXCUSE

You've just harnessed the power of static electricity to hold the pepper on to the rubbed section of plastic lid. You obviously know why this happened, but will your fellow explorers?

TAKE CARE!

Make sure you give the flannel a good rub; it's virtually impossible to rub it too much, but not doing it enough will lessen the effect.

CRAZY CEREAL

Every science book needs at least one experiment that lets people go crazy over breakfast. And this one provides you with the least likely answer when your mother tells you to hurry up with your cereal: 'But Mum, the cereal just won't stay on my spoon!' You'd better be able to come up with a bit more than that, though, so read on until you feel happy enough with your excuse.

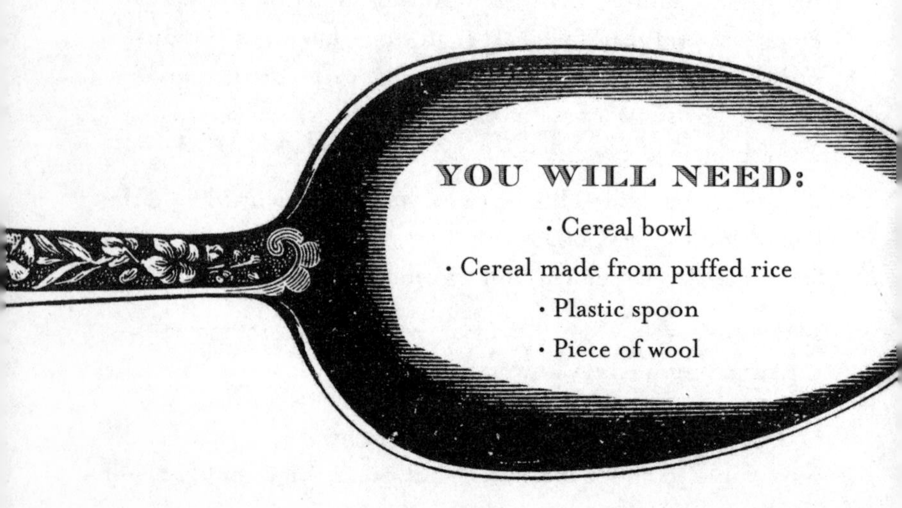

YOU WILL NEED:

- Cereal bowl
- Cereal made from puffed rice
- Plastic spoon
- Piece of wool

METHOD

1. Fill the bowl with the cereal as normal, but do not add milk.

2. Rub the spoon on the wool for about 15 seconds.

3. Hold the spoon over the bowl.

4. The bits of puffed rice will start to dance and bounce around, zinging off the spoon and all over the table.

THE SCIENTIFIC EXCUSE

Might this one really be called the oldest trick in the book? That's up to you. Of course, we're back in static electricity territory. Rubbing the spoon on the wool gave it a negative electrical charge. This charge, in turn, attracted the bits of puffed rice until they touched the spoon. That's when they transferred some of their electrons, and once their charge was the same as the spoon's, they dropped off.

TAKE CARE!

Make sure you clear up after yourself. Even smart-alec scientists know when to be polite.

MICROWAVE SOAP

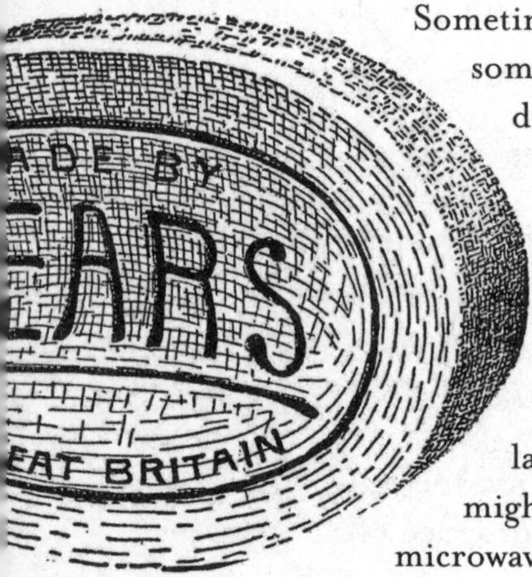

Sometimes you can learn a lot about something by looking at it from a different angle. This way-out experiment seems to be all about making merry with an 'explosive' bar of soap. There's no denying that the result of this demonstration will trigger 'oohs', 'aahs' and hoots of laughter. But along the way, you might finally get an idea of just how microwaves really work.

YOU WILL NEED

- Bar of soap
- Piece of paper towel
- Microwave oven

METHOD

1. Put the soap on the paper towel and put both in the centre of the microwave.

2. Cook the soap for two minutes on the highest setting.

3. Allow the soap to cool for one minute before touching it.

4. The cooled soap will be remarkably puffy but still rigid.

THE SCIENTIFIC EXCUSE

The key to this experiment is in the water, even though that wasn't one of the ingredients. All soap contains water, both in the chemical compound of the soap itself and also in the form of water vapour inside tiny air bubbles within the soap (the number of air bubbles can vary between soaps). The heat from the microwave causes the trapped water to vaporise and the trapped air to expand. Meanwhile, the soap has become much softer so that it puffs out because of all the expansion within it.

TAKE CARE!

The normal precautions about using a microwave apply to this experiment, which mean that it must be carried out by an adult. Be careful to let the 'cooked' soap cool before you touch it.

The Final Frontier

It's time to get in a little over your head with some exploits and experiments that harness, or at least track, the stars above. You might never get a chance to use a sand wedge on the Moon, as Alan Shepard did, but you can put the Sun to work. While you're at it, there's an opportunity to create a mini solar system and to make a slow getaway ... a solar-powered one, naturally.

THE LARGEST SAND TRAP

It has been noted many times that people can become obsessive about golf. Is it because for just one putt – or maybe even for an entire hole – a hacker can feel like the equal of Tiger Woods or Jack Nicklaus? Or maybe because striding around an 18-hole course gives stressed-out executives their only chance to get away from it all, if only for a few hours?

Hollywood star Samuel L. Jackson insists on time off for rounds of golf when he negotiates film contracts. *Golf Punk* website and magazine caters to a new generation of golfers who prefer street fashion and crazed antics to blazers and club dinners. Some golfers have taken to the streets – literally – to enjoy rounds of urban golf, with bins taking the place of holes and 'fairways' narrowed by the buildings on either side.

But the first prize for craziest golf stunt ever must go to NASA astronaut Alan Shepard, on 6 February 1971. Shepard had been America's 'first man in space' in May 1961, but his achievement had been eclipsed by John Glenn's first orbital flight nine months later. The 1971 lunar mission, Apollo 14, would restore the shine of his reputation.

A lot was riding on the outcome of Apollo 14. The previous mission, Apollo 13, had nearly ended in disaster after a series of malfunctions. The Americans needed a success – and a scientific success for that matter – to justify the enormous sums being spent on the space programme. In the event, Shepard and his colleague Edgar Mitchell surpassed the tough goals set: they ventured further from the landing craft than any previous mission, spent more than nine hours exploring the surface and then collected nearly 50 kg of Moon rock.

So perhaps no one back at mission control was too cross when Shepard took a break just before blasting off from the Moon's surface. He pulled out a modified (collapsible) six iron and two golf balls that he had sneaked inside his spacesuit. Then he whacked one after another into the lunar distance – 'for miles and miles and miles', as he noted aloud into his radio transmitter.

Follow That Star!

This book isn't all about quick-result, high-drama exploits. And if you're thinking of the heavens, you need to operate on a different timescale – one that uses light years and not minutes to get results. This experiment reveals an exciting and beautiful result but it takes its time to achieve it: one Earth year, to be precise. And the star that you'll be following through that time? Why, it's our Sun, of course.

YOU WILL NEED

• One nearby star (the Sun is ideal)

• A flat area (5 m square or more): smooth lawn is best, but a paved area will also do

• A strong pole at least 80 cm tall (or an existing pole secured in the ground and at least 50 cm tall)

• Mallet

• Spirit level or plumb line

• Tent pegs or paint

• Chalk or string (at least 10 m)

METHOD

1. Drive the pole into the ground so that it is secure and stands at least 50 cm tall.

2. Use a spirit level or plumb line to ensure that the pole stands straight.

3. Decide on what you will be using for the next year as

markers on the flat area. Chalk is no good because it will wash away, so use either tent pegs (on lawn) or paint (on paved areas).

4. Note where the shadow of the pole appears at noon. Mark it with your permanent marker (peg or paint).

5. Continue to mark the noontime tip of the shadow at regular intervals – every week or two – for the next year.

6. Don't worry if you have a run of cloudy weather with no visible shadow – just resume when you can.

7. Draw the shape made by connecting the markers – with chalk (on paving) or by threading string through the tent pegs.

8. You should have produced a beautiful figure of eight.

— THE SCIENTIFIC EXCUSE —

It might have taken you a year, but you will have managed to demonstrate two important features of our planet. The long, figure-eight design that you plotted is called an analemma. The north-south difference of the points demonstrates the tilt of the Earth: the Sun (like its shadow) appears to move up and down through the seasons. The east-west differences are evidence that the Earth's orbit is an ellipse, and not a pure circle.

TAKE CARE!

You will have to ignore the change in clocks through the year for this one. In other words, if you begin the experiment in the winter (before clocks go ahead), take your reading at 1 pm (the same as noon, before the clocks went ahead) throughout the summer.

SOLAR-POWERED OVEN

These days everyone is talking about conservation and saving the environment. Recycling, alternative and renewable energy sources – these are all very good ideas. But what can we do as individuals to change our ways and help out? How about turning to our nearest star to lend a hand in the kitchen? But take note: you have to eat a pizza before starting out on this experiment.

YOU WILL NEED

- A large, recycled pizza box (the sort you get when you order a large takeaway pizza)
- Aluminium foil
- Black sugar paper
- Ruler
- Sheet of sturdy clear plastic (preferably laminated)
- Non-toxic glue
- Sticky tape
- Scissors
- Felt-tip pen

METHOD

1. Draw a border on the top flap of the pizza box, about 3 cm in from the edge.
2. Carefully cut along three of those four lines, leaving the line along the 'hinge' of the box alone. Open and close several times to form a crease.

3. Cut a piece of foil the same size as this flap; glue it to the inside (lower) edge of the flap.

4. Measure and cut a piece of plastic just a bit larger than the opening; tape this plastic to the underside of the box top, making sure it covers the cut-out hole to form a complete air seal. At this point, the box top should have this new plastic seal on the underside; the foil-backed flap opens up from this plastic layer.

5. Cut a second piece of foil and glue it to the bottom (inside) of the pizza box.

6. Cut some of the black sugar paper to fit this same base; tape it to the foil on the base.

7. Aim the box so that it opens towards the Sun.

8. Prop the flap open – but with the box top shut – to get the oven working.

9. You can cook all sorts of things – muffins, crumpets and maybe even more pizza – provided that they don't protrude higher than the box top.

— THE SCIENTIFIC EXCUSE —

The oven works by focusing the Sun's rays towards the food tray. The foil increases the number of rays entering the oven. They can pass through the plastic, but the heat stays inside. The result – evidence you can eat.

TAKE CARE!

Bear in mind that you have actually constructed an oven, which can reach temperatures of up to 135 °C. But it takes its time to reach that temperature. Leave half an hour for the oven to preheat, and then figure on cooking things for twice as long as you would in a normal oven.

Bending Light

What did Einstein say about light beams bending as they travel through space? He did mention drinks bottles, didn't he? Even if he didn't he should have, as you'll be able to show with this wonderful head-scratching display of wizardry. Make sure that when you've got everything prepared for the experiment, you really do turn off the lights in the room so that you'll get the best from this other-worldly demonstration.

YOU WILL NEED

- Empty 1-litre drinks bottle
- Torch (with its shining end about the same diameter as the base of the bottle)
- Aluminium foil
- Sticky tape
- Sink
- Audience

METHOD

1. Wrap a layer of foil around the outside of the bottle, leaving the top and bottom uncovered.

2. Secure this foil with tape if it seems a bit loose.

3. Fill the bottle with water and then tighten the cap. At this point you should make sure you are near the sink.

4. Position the torch snugly up to the base of the bottle.

5. Ask someone to turn off the room light and then turn the torch on.

6. Tilt the bottle down so that it's a little higher than parallel with the floor, and unscrew the cap.

7. Watch as the watery light beam escapes from the bottle and bends downwards into the sink.

— THE SCIENTIFIC EXCUSE —

This basic experiment demonstrates fibre optics, the process that allows photons (particles of light) to travel down a curving tube in the same way that a bobsleigh zooms through the curves of its run. How much of the light stays inside the tube and how much is reflected back depends on the angle of the light as it enters the tube. In this demonstration, the light from the torch travels along the tube because it reflects back from the foil. Better yet, it stays part of the 'tube' (this time reflecting back from the edge of the water itself) as it flows – and curves – down from the opening of the bottle.

TAKE CARE!

There's no danger in this experiment, but it works best with a powerful torch that is about the same size as the base of the bottle. That way, you get the most dramatic flow of photons with little or no light 'leakage' at the torch end of the bottle.

The Revenge of Icarus

And it is oh, so sweet! The ancient Greek lad got a little too close to the Sun, which led to his comeuppance. How about turning things on their head and actually using the same Sun to get a lift? Is this all just a lot of hot air? You decide. Just make sure you've chosen a calm, sunny day to try this out.

YOU WILL NEED

- Large black bin bag
- Twister seal
- String
- Paperclip
- Crayons
- Paper
- Scissors

METHOD

1. First of all, construct a fake £5 note (or other important document) by cutting a piece of paper and then colouring it in with crayon.

2. Hold the bin bag open with both hands and then whoosh it round until it is almost full of air.

3. Twist the top and seal it with string or a twister seal.

4. Tie a 2 m length of string to the knotted end of the bag and then clip the '£5 note' to the other end of the string.

5. Just wait. Depending on how warm it is outside, the bag will start to rise and then float up, up – and away, taking your money with it?

— THE SCIENTIFIC EXCUSE —

The Sun does all the work here. Its rays provide us with warmth anyway, but the black colour of the rubbish bag means that it absorbs even more heat. This warms up the air inside, which expands as it warms. But because the same amount of air now takes up more space, it loses density. And because the air inside is less dense than the outside air, it rises.

TAKE CARE!

This is a low- to no-risk experiment *unless* you decide to abandon the 'fake document' idea and use something that really is valuable.

Impact Craters

Everyone loves to make a little mess now and then. How about channelling this urge into something scientific and productive? This experiment reveals its evidence in one of the most babyish of settings – a sandpit. But what it tells us can help shed light on some of the mysteries of the heavens. Make a record of this experiment and you'll be on your way to mastering the scientific method.

YOU WILL NEED:

- Sandpit (the finer the sand, the better)
- Set of kitchen scales
- Notebook and pencil
- Ruler
- Five objects of different sizes and weights (for example a golf ball, a tennis ball, a marble and two stones)

METHOD

1. Smooth out the surface of the sandpit. This will represent the surface of Mars or the Moon.

2. Weigh each of the five objects and record that information.

3. Stand over the sandpit, holding the first object at arm's length, and let it drop down to the sand.

4. Measure the depth and width of the crater it produces.

5. Record this information.

6. Continue steps 3 ,4 and 5 for the other objects.

THE SCIENTIFIC EXCUSE

Each of the 'bowls' that forms is an impact crater. Bits of rock and ice are constantly flying towards Earth from outer space. Most of them burn up harmlessly in the atmosphere. But Mars has a much thinner atmosphere, and the Moon has none at all, so now you can see why their surfaces are so scarred. Don't lose your records: you can now draw conclusions about size, weight and the craters that are formed as a result. Next stop – the International Space Station.

TAKE CARE!

Remember that this is in the cause of science, and not Olympic shot-put trials. Simply *dropping* each object means that they fall at the same rate. If you started throwing some of them, you would be introducing what scientists call a variable – something that is different with each go. You don't want to lose your reputation as a precise scientist – you might lose millions in funding for your next project!

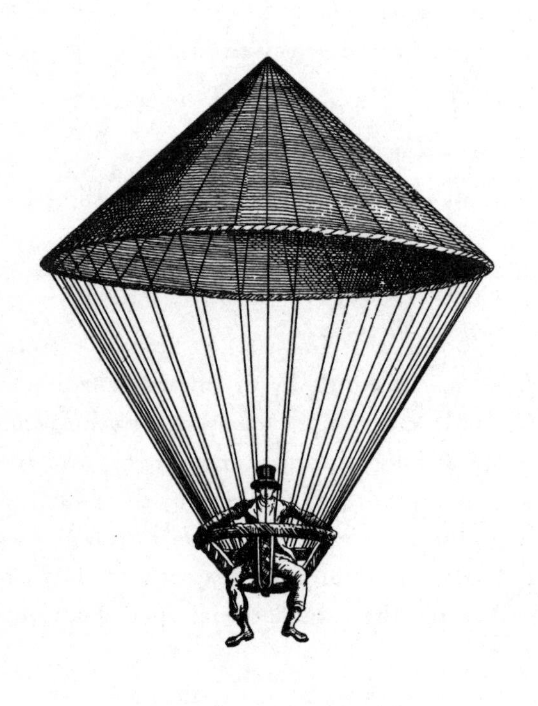

Something in the Air

Politicians don't have a monopoly on hot air. Scientists can give them a run for their money; plus they can do a lot with cool air, cold air and all sorts of other gases. Test-drive some of the following suggestions and you'll see why. Sure, these experiments carry an element of risk and edginess. But at least they're not likely to land you in a Russian labour camp – the fate that befell the young German pilot whose day trip to Red Square turned out to be one-way.

THE RED SQUARE EXPRESS

It came as something of a surprise to the wider world to learn that one May evening in 1987, a private aircraft from Germany lazily circled central Moscow, buzzed Lenin's tomb and taxied to a halt by the Kremlin walls. What made the story even more unlikely was the fact that the pilot, Mathias Rust from West Germany, was only 19 years old.

It is true that the Soviet Union in 1987 was a less menacing place than it had been during the dark decades of Stalin's rule. Led by a young and vigorous leader, Mikhail Gorbachev, the country was slowly ridding itself of the worst excesses of communism. People were freer than they had been for decades to express themselves, and even to offer some mild criticism of the government.

On the other hand, the Soviet Union still prided itself on its might and weaponry. Each May Day saw ranks of Soviet troops and weapons parade through Red Square – the very spot that Rust chose as a landing runway. And the Soviets were not shy about defending their territory from intrusion: in 1983, Soviet fighters had shot down Korean Airlines flight 007, which had strayed into Soviet airspace. All 269 people on board had been killed.

Rust had succeeded by sticking to a simple plan. He had flown his single-engine Cessna out of Helsinki, telling controllers that he was going to Stockholm. Then he turned east, switched off radio communication and carried on across 700 km of the most fiercely-defended airspace on Earth. He passed through various zones of Soviet air defence, either by flying too low for radar detection or because his plane was confused with Soviet training aircraft. Some reports even suggest that he landed along the way, throwing pursuers off the trail.

A small group of bewildered pedestrians surrounded Rust's plane when he did come to a halt by the Kremlin. Soon afterwards, he was arrested. Rust went on trial in September 1987 and was sentenced to four years in a labour camp. He served just over a year, and was released on parole before being sent back to West Germany. His idea had been to promote peace through his flight. In the end, he was simply relieved that he hadn't caused a Third World War.

Don't Pressure Me

Some of the most astounding scientific displays can be performed with a minimum of fuss and special equipment. This counter-top performance demonstrates one of the most important unseen forces around us – air pressure. It might be invisible, but it certainly does its bit to make it hard to p-u--, p-u-u-u-l, *pull* this dratted carrier bag out.

YOU WILL NEED

- Metal or stout ceramic mixing bowl (25–30 cm diameter)
- Strong elastic band (the sort used by postmen should do fine)
- Plastic carrier bag (check that there are no ventilation holes)

METHOD

1. Line the inside of the bowl with the carrier bag, feeding any surplus up and around the rim.

2. Secure the plastic overlap to the bowl with the elastic band.

3. Put the bowl-bag combination on a table or counter and ask a friend to pull the bag out of the bowl by holding it at its centre.

4. Your friend will find it very hard – and maybe impossible – to make the bag budge.

— THE SCIENTIFIC EXCUSE —

To understand this experiment, you must think a bit about air pressure. By securing the bag to the outside of the bowl, you have trapped some air between the outside of the bag and the inside of the bowl. By pulling up on the bag you are trying to increase the space (volume), and that is when Boyle's Law comes into play. If the volume of gas (in this case, air) in a closed system increases, its pressure decreases. That is exactly what is happening inside the bowl and under the bag. The air pressure *outside* the plastic bag, however, retains its strength – forcing the air hard against any increase in volume.

The German scientist Otto von Guericke took this demonstration even further in 1650, by sucking all of the air from a globe created from two joined copper hemispheres. With no air pressure to help them from inside the globe, teams of horses connected to each half failed to pull the hemispheres apart.

TAKE CARE!

Make sure that you line the inside of the bowl as much as possible in order to help create the largest possible air pressure difference. Try to get the best match that you can between the sizes of the bag and the bowl.

– The – Sliding Tumbler

Or should that be the tumbling slider? No – that will have to wait for another volume, like 'The Tale of the Giant Rat of Sumatra' (which Sherlock Holmes never got round to telling Dr Watson). At any rate there's no real need to look any further, because this little gem provides that subtle hint of magic that turns experiments into lasting memories. Try it a few times – you'll certainly warm to it.

YOU WILL NEED

- Plastic-coated tray
- Glass tumbler
- Candle
- Matches
- Water
- Several small paperback books

METHOD

1. Slide a book under a narrow end of the tray.

2. Wet the rim of the tumbler and put it upside down on the tray near the top.

3. The tumbler will stay where you set it down.

4. Take a lighted candle and put it close to, but not touching, one side of the tumbler.

5. The tumbler will start to slide slowly down the slope of the tray.

THE SCIENTIFIC EXCUSE

The water round the rim makes an air seal inside the tumbler. Then the candle warms the air inside, causing it to expand. The expanding air pushes the tumbler slightly up, so that it now 'floats' over the tray with just a seal of water to help it slide. The rest is there to be seen.

TAKE CARE!

It's worth trying this experiment out a few times on your own to work out exactly how many paperbacks produce a decent slope – but not so steep that the tumbler slides down without help from the candle.

MATCH ALERT!

This experiment involves the use of matches and should be conducted only by a responsible adult.

Bull-roarer

The Aboriginal people have lived in Australia for more than 40,000 years. They have a special relationship with the land, and many natural features – mountains, river beds and even large rocks – are sacred to them. These people have invented special musical instruments to sound alarms when people stray too close to these sacred sites. Most people have heard the mournful didgeridoo. This experiment helps you make another weird-sounding Aboriginal instrument – the bull-roarer. If you get it right, you'll produce sounds that seem to come from another world.

YOU WILL NEED

- Eight thick elastic bands
- A piece of card (at least 35 cm x 25 cm)
- Scissors
- Pencil
- Sticky tape

METHOD

1. Measure and draw an oval shape (30 cm wide by 7 cm deep) at one end of the card.

2. Cut out this shape and then use it to trace another, exactly like it, from the card. Cut this second oval out.

3. Repeat step 2 to make a third oval.

4. Fit the ovals over each other snugly and tape them in place.

5. Use the point of the scissors to poke a hole through all three layers, about 2–3 cm from one of the narrow ends.

6. Make a chain from the elastic bands by looping them together.

7. Feed one end of the chain slightly through the hole, and then loop the other end over and attach it to this end. (This makes the chain into a large loop, with the cardboard oval attached.)

8. Find a good-sized open space and whirl the bull-roarer round and round your head: you'll hear the eerie noise in a few seconds.

9. To get an even more dramatic result, try whirling the bull-roarer round inside a tunnel (like a pedestrian subway).

THE SCIENTIFIC EXCUSE

As you whirl the oval blades round your head, they begin to spin rapidly. This spinning causes the air molecules around them to vibrate. And since what we process as sound is really vibrating air, the result is an uncanny noise.

TAKE CARE!

Scissors must be handled with great care. Make sure an adult pokes the hole through the cardboard ovals.

Ping-pong Pump

You're in the middle of a tough ping-pong game with your best friend. The winner will have bragging rights for the next year. It's tough – the lead, like the game itself, is going back and forth. Forehand, backhand return, backhand, lob ... smash! The ball nicks the table for a winner and you've got match point – except your opponent picks up the ball. He shows you a dent where it caught the edge of the table: 'Too bad. That's our last ball. Looks like we'll have to settle this another time.' He's probably right. Your moment of glory will have to wait.

Or will it?

YOU WILL NEED

- A dented table-tennis ball (see note below about dents and cracks)
- Hot water (the hottest from the tap should do)
- Small mixing bowl or jam jar
- Empty plastic drinks bottle (with lid off)

METHOD

1. Fill the bowl or jar two-thirds full with hot water.

2. Drop the ball into the water.

3. Hold the drinks bottle upside down with the opening just touching the floating ball.

4. Carefully push the bottle down so the ball goes underwater.

5. Keep the bottle and ball in place until the dent pops out (this usually takes less than a minute).

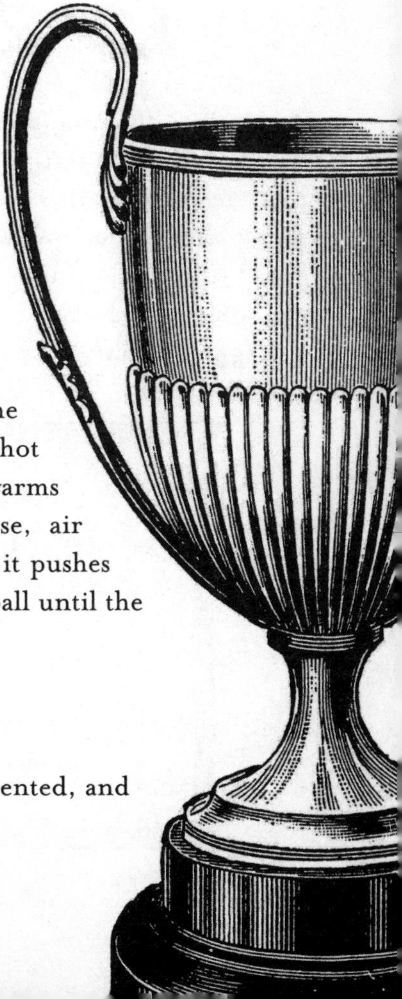

THE SCIENTIFIC EXCUSE

You've saved the day – and your chance to win the match – with some basic science. The water is not hot enough to damage the ball but it warms the air inside it. And, of course, air expands as it warms. In this case, it pushes outward against the inside of the ball until the outside pops back into shape.

TAKE CARE!

This only works on balls that are dented, and not pierced or cracked.

– The – Hankie Dam

Admit it – dumping a glass of cold water over someone's head is funny. In fact, without glasses of water and custard pies, the world would have been robbed of some classic slap-stick routines. But how funny is it for the person under the glass? And would you be willing to try it over the head of, say, your elderly aunt? Read on to see how you might manage this stunt – and survive.

You Will Need:

- Drinking glass
- Water
- Handkerchief
- Elderly aunt (optional)

METHOD

1. Fill the glass with water.

2. Spread the handkerchief over the top of the glass, holding it tightly in place.

3. Find an aunt or anyone else prepared to take part.

4. Still holding the handkerchief in place, flip the glass and hold it upside down over the person's head.

5. The water won't spill.

THE SCIENTIFIC EXCUSE

Two forces are at work here – air pressure and surface tension. The air is pressing up on the fabric of the handkerchief. This fabric, however, has become saturated with water and therefore has strong surface tension. The air is pushing up but meeting a barrier in the form of surface tension. Provided the glass is held exactly upside down, the air pressing up and the water pressing down are kept apart because of the surface tension. But if you hold the glass at an angle, or let some slack into the hankie, more water can press down in places – and air can press up in others. That can mean only one thing: a spill.

TAKE CARE!

Just to be on the safe side – and to help you gain confidence – you can practise this a few times over a sink or basin.

Bottling Out

This experiment works best if you perform it as a sort of competition. The challenge is simple: who will be the first to blow up a balloon inside a bottle? Find one of your most competitive friends and lay down the challenge. But to make sure you come out on top, complete steps 1 to 4 before you invite your friend to compete.

• YOU WILL NEED •

- 2 balloons
- 2 empty plastic fizzy drinks bottles (1.5 litre capacity)
- Sharp knife or scissors
- Volunteer

METHOD

1. Cut a 1 cm hole in the bottom of one bottle.

2. Stretch a balloon across the mouth of each bottle.

3. Use your little finger to push each balloon gently inside its bottle (keeping it attached to the mouth).

4. You should now have a balloon hanging loosely inside each bottle but still attached to the top.

5. Ask your friend to race you to see who can blow up a balloon fastest. (Hand your friend the bottle with no hole in it, and make sure they don't see the hole in yours.)

6. Blow up the balloons.

7. Yours – inside the bottle with the hole – should blow up pretty easily. Your friend, though, will be turning all sorts of colours and failing to get anywhere!

THE SCIENTIFIC EXCUSE

Like so many of the best experiments, especially the ones that leave people steaming, this has a simple explanation. When you blow up a balloon, it expands to make room for this extra air. It does this by pushing aside the air molecules that surround it. The air molecules in the 'holed-out' bottle get pushed out of the hole, which amounts to the same thing. The air molecules in the other bottle, though, have nowhere to go, so there's no room for the balloon to grow.

TAKE CARE!

An adult should cut the hole in the drinks bottle, because the plastic is strong and you need an equally strong, steady hand to cut through it.

Air Block!

Here's a chance to mix sleight of hand with science – and in the process, a chance to impress people and maybe even win a bet. What's at stake, apart from anything you might wager with a sceptical friend, is whether you can fill the bottle with water. It's nearly there to begin with, so it shouldn't be hard. 'You sir, I bet you can't manage it – and that I can.'

YOU WILL NEED

- 1-litre bottle (glass or plastic)
- Funnel (with a spout narrow enough to fit into the bottle)
- Plastic jug
- Modelling clay
- Water
- Small pin
- Volunteer
- Witnesses

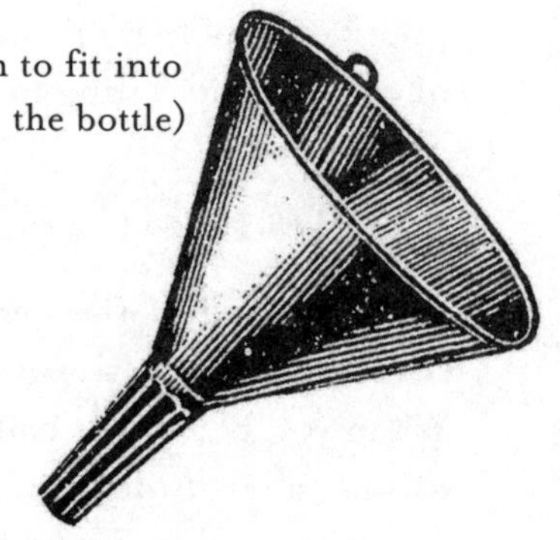

METHOD

1. Fill the bottle two-thirds of the way up with water.

2. Feed the funnel into the opening of the bottle. There should be a good 5–8 cm between the bottom of the funnel spout and the water level inside the bottle.

3. Secure the funnel tightly in place with modelling clay.

4. Fill the plastic jug with water and ask your victim to fill the bottle using that water.

5. The funnel should fill up easily enough, but your friend will find it impossible to get any of the water from there into the bottle.

6. Here's the tricky bit: distract your friend and the witnesses for a split second and stick the pin into the clay (so it goes between the spout and the funnel inside the bottle).

7. Slowly pour some more water in with a flourish, and while people are watching your pouring hand, pull out the pin.

8. The water should pour right in.

— THE SCIENTIFIC EXCUSE —

It doesn't look as though anything is inside that bottle above the water, but in fact there is something that's pretty strong – air. That air has filled the space, and with no way for it to escape (you've seen to that with the clay), it will stop any water from entering. The pin prick allows air to escape and water to flow in.

TAKE CARE!

The pin, of course, is the secret weapon. Have it hidden somewhere – maybe stick it on your belt loop so you can pull it out easily when everyone else is distracted. Use a pin with a head large enough to be felt once you've stuck it in: you don't want to have to search for it.

Weight of the World

Air, like water or ice-cream or mud, has both mass and weight. In other words, it can be weighed just like those other substances. The total mass of the air surrounding us – what the scale would read if you weighed it – is 5,140,000,000,000,000,000 kg. And if your friends find it hard to believe that air could weigh *anything*, try out this experiment on them.

YOU WILL NEED

- 2 balloons
- String
- Scissors
- 1 m rule
- Sticky tape
- Pin

METHOD

1. Cut off two strips of sticky tape, each exactly 2 cm long; keep these handy.

2. Cut two 40-cm lengths of string.

3. Blow up both balloons until they are exactly the same size and tie them off.

4. Tie one end of each length of string to the balloon knots.

5. Use the saved sticky tape to attach the other end of the strings to the measuring stick – one at each end.

6. Hold one arm outstretched and balance the stick (with balloons at each end) on an outstretched finger.

7. Now pop one of the balloons with the pin.

8. The other balloon will drop and the stick will fall off.

THE SCIENTIFIC EXCUSE

Blowing up the balloons fills them with air – we all know that. But it also means that the inflated balloon weighs more because of this extra air that has squeezed in. Popping a balloon allows the air to escape, and the balloon loses the extra weight.

TAKE CARE!

The risks to you? Well, there are two. The first is seeming like too much of a know-it-all before you get going (so maybe you'd better go easy with that 'kg total' figure). The second is seeming like too much of a know-it-all after you've done this simple demonstration. Wipe that smirk off your face – now!

Balloon Hang Ten

This daring experiment has come to us all the way from Australia. Maybe after you've mastered it, you'll see why the Aussies are among the world's best surfers. From Fremantle to Bondi Beach, those daring denizens of Down Under have proved their skills balancing on surfboards in the biggest of waves. Is it possible that they practise in their sitting rooms in the off-season? See what you think after you've had some practice yourself.

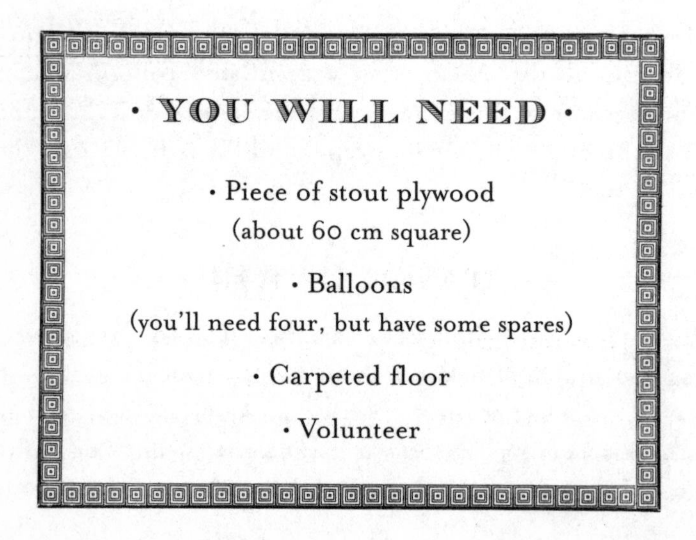

· YOU WILL NEED ·

· Piece of stout plywood (about 60 cm square)

· Balloons (you'll need four, but have some spares)

· Carpeted floor

· Volunteer

METHOD

1. Half-fill four balloons and tie them off.

2. Put the four balloons under the piece of plywood on the floor.

3. Have your volunteer hold the plywood steady as you step onto it.

4. Wait for one of the balloons to pop – none of them should!

THE SCIENTIFIC EXCUSE

This is one of those experiments that recalls the elephant's footprint: it's shallow because the wide foot spreads the weight and lessens the pressure at any one point. The same holds true here. When you stand on the wood, your weight exerts downward pressure. But the balloons are only half full, so they can flatten and widen out – increasing the exposed area and reducing the force. Surprisingly, they reach a sort of balance point at which they've flattened enough to support your weight – and maybe even have room for more.

TAKE CARE!

Don't underestimate the surfing angle here: it really is easy to lose your balance when you're on such a set-up. Keep your arms outstretched – like a tightrope walker – to help you keep upright. And it might be sensible to remove breakables from the experiment zone.

HEAVY READING

Your father is the sort who gets everything just so – cushions plumped, reading lamp positioned, footstool in place – before he settles down to read his Sunday paper. So will he take kindly to seeing that today's edition begins on page 3? How much happier will he be to see you spreading the paper on the table and attacking it with a ruler? It looks as though you've got some explaining to do. Over to you, professor.

YOU WILL NEED

- A full sheet of newspaper (that is, numbered pages such as pages 1 and 2, and the last two pages)
- Wooden ruler
- Level table or large counter

METHOD

1. Spread the newspaper sheet flat (with the fold pointing down) on the table or counter.

2. Make sure the paper is more or less flush with the edge of the table or counter.

3. Slide the ruler under the paper until only about 10 cm of it still sticks out from the table.

4. Slap your hand down on this extended bit of ruler.

5. Instead of being launched into space, the paper will stay pretty firmly in place.

THE SCIENTIFIC EXCUSE

Scientists use the word 'counter-intuitive' to describe events that seem to contradict our expectations. And we would certainly expect to be able to send that light piece of paper into orbit. But this experiment demonstrates how we'd be ignoring something very important – air pressure. The force of the air pushing down over a fairly large area – all of the newspaper sheet – is greater than the force of the ruler pushing up.

TAKE CARE!

This experiment works best with the largest newspaper sheet possible. What you really want is part of a broadsheet newspaper (like the *Sunday Times* in the UK or the *Sydney Morning Herald* in Australia), rather than a smaller tabloid newspaper.

Straw Rocket

That means drinking straw, and not 'scarecrow' straw, by the way. But even a drinking straw rocket might seem a little far-fetched unless you can figure out how to harness Newtonian physics. After all, if Sir Isaac could start to theorise about gravity after being hit by an apple, just think what he might have come up with had he had plastic straws at his disposal.

Hint: you might find it easier to locate straws by popping into a nearby branch of a famous burger chain or high-street coffee chain. Of course, you'll want to buy something to keep your conscience clear.

You Will Need

- 2 plastic drinking straws (note: they must have different diameters)
- 200 ml squeezable fruit drinks bottle (such as Robinsons Fruit Shoot bottle)
- Modelling clay
- Small piece of cotton wool

Method

1. Make sure the drinks bottle has been rinsed thoroughly and has dripped dry.

2. Feed the narrower straw into the opening of the drink

bottle and plug any gaps with modelling clay.

3. Plunge one end of the wider straw into some more modelling clay to make a plug. Trim any excess (while making sure it still works as a plug).

4. Stick a small wad of cotton wool over the outside of the clay plug. This will make the end of the rocket less dangerous when it is launched.

5. Slide the wider straw down over the narrower one. Let it rest gently (do not jam it in) and hold the bottle upright so that the straw points upwards.

6. Have your friends perform the countdown, and when they reach 'zero' squeeze the bottle hard.

7. The outer straw will be launched into space.

— THE SCIENTIFIC EXCUSE —

We did say Newtonian physics was at work here – his Third Law of Motion to be precise (that for every action there is an opposite and equal reaction). The action is the air being forced through the smaller straw. It hits the plug at the end of the wider straw and triggers the reaction – focusing on the small plug so that the force is channelled into launching the straw rocket.

TAKE CARE!

Make sure you point the rocket upwards, and not at anyone near you. Even a flying straw can cause an injury.

THE CAN CAN-CAN

There's a wonderful scene in the film *Jaws*, where the tough shark-hunter Quint finishes his beer and crushes the can in his bare hand. You can perform a variation of this feat – with even more drama – by demonstrating some of the wonderful powers of water. You might not get the shark, but you'll impress some of the nearest landlubbers.

YOU WILL NEED

- Empty 330 ml drinks can
- Large saucepan
- Water
- Frying pan
- Oven mitt or tongs
- Stove
- Tablespoon

METHOD

1. Heat the frying pan on the stove until it is hot enough to use for cooking.

2. Meanwhile, fill the saucepan to a depth of about 3–4 cm with cold water.

3. Carefully pour a tablespoon (15 ml) of water into the can.

4. Use the mitt or tongs to place the can on the frying pan.

5. Leave the can there for about 10 seconds and then use the mitts or tongs to transfer it to the saucepan.

6. Place the can in the saucepan upside down.

7. The can should collapse rapidly with a loud crunch.

THE SCIENTIFIC EXCUSE

Lots of interesting experiments demonstrate how air expands when it is heated up. This experiment gives that a little twist by heating up the small amount of water in the tin until it boils. The tablespoon of water turns into water vapour, a gas that drives the air out of the tin. By plunging the tin into the cold water, the temperature plummets and the vapour turns back to liquid water (in a process called condensation). But turning back to liquid means that the water takes up far less space ... and as all of the air was driven out earlier, there's nothing to stop the pressure of the saucepan water pushing in on it.

TAKE CARE!

Because of the stove, hot pan and boiling water, this experiment must be carried out by an adult.

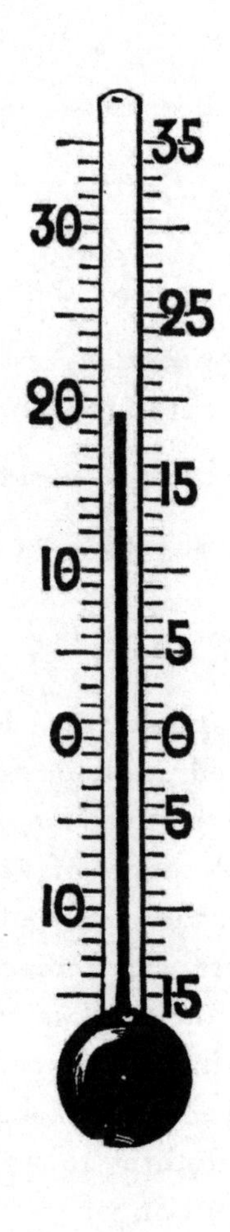
35
30
25
20
15
10
5
0
0
5
10
15

Climate Change

We all love to create a good atmosphere. But read on, and you'll see how you can do just that – literally. Whether it's calling the shots in cloud production or feeding funny fumes into a family function, if you try out the experiments in this chapter you'll feel on top of the world. Plus you'll see how being reassuring can be just as irresponsible as urging people to go wild. Just ask the Met Office about the 1987 hurricane ... if you dare.

The Hurricane That Wasn't

The weather is one of the few topics where the British abandon their love of understatement. A few snowflakes blowing over some high ground become a 'blizzard'. A north wind immediately becomes 'polar'. Summer sun that would pass as early spring in Portugal or Peru becomes a 'scorcher'. Can you blame a leading Met Office forecaster for trying to calm viewers who might be expecting a hurricane?

That is exactly what Michael Fish tried to do in his evening forecast on 15 October 1987. Throughout the afternoon, a powerful storm was developing in the Bay of Biscay along the Atlantic coast of France. And it was moving north towards the southern half of Great Britain.

Things get a bit peculiar at this point. It's likely that 99 per cent of viewers who saw the BBC TV forecast that night will remember this comment from Mr Fish:

> Earlier on today apparently a lady rang the BBC and said she heard that there was a

> hurricane on the way. Well don't worry if you're watching, there isn't.

They will also remember the storm that ripped across the country that night after most people had gone to bed, with gusts reaching nearly 200 km/h. Sixteen people died because of storm damage. Buildings broke, collapsed or were very nearly blown away. Ships ran aground along the Channel coast, where flooding also became a problem. An estimated 15 million trees were uprooted in the storm. Weather experts considered it to be the worst storm in southern Britain since 1703.

Poor Michael Fish immediately found himself linked to the most misguided bit of reassurance since Icarus boasted about his wings. But did he have a point?

He could use two lines of defence. First, strictly speaking, the storm of 1987 was not a hurricane, since it did not originate in the tropics and did not have sustained wind speeds of 125 km/h. Second, he did actually advise viewers to 'batten down the hatches because there's some really stormy weather on the way'. The 'don't worry' quote actually referred to a news item earlier in the bulletin about a hurricane near Florida. That hurricane, of course, never reached Britain. But the damage had already been done.

Create a Cold Front

We all know what it's like to listen out for a weather forecast and then be disappointed when it turns out wrong and our plans go down the drain. Wouldn't it be lovely to be able to predict when it will turn cloudy – and even when it will begin to rain – right down to the last second? Actually, it might be easier than you think to pull off that trick.

YOU WILL NEED

- Wide-mouthed glass jar
- Hot water
- High-rimmed metal baking tray
- Tray of ice cubes
- Torch

METHOD

1. Place the jar on a counter and fill it to a depth of about 3 cm with hot water.

2. Empty the ice cubes into the baking tray and set the tray on top of the jar.

3. Turn off the main room light.

4. Shine the torch into the jar to reveal a foggy cloud forming inside.

— THE SCIENTIFIC EXCUSE —

You've created a mini weather system in this experiment. The hot water at the base of the jar warms the air, which then rises. This air has also absorbed some of the water in the form of tiny droplets. At the top, though, it hits the cold base of the baking tray. The humid air cools rapidly, forming a cloud – just as your warm, humid breath does when you breathe out on a cold day. Look a bit closer and you'll see that some drops of water are forming on the bottom of the tray and are about to drop off. That is just what happens when raindrops form.

TAKE CARE!

Make sure you use hot, but not boiling, water in this experiment. Boiling water could cause the glass to crack, which could in turn lead to cuts and scalds.

Being Particular

Just about every experiment in this book involves your being irresponsible – or at least your being accused of irresponsibility. Why not try one that looks at the irresponsibility of others – how does measuring air pollution sound, for a start? This experiment relies on some basic carpentry, patience and a sense of outrage for its full effect.

YOU WILL NEED

- 150 cm-long broom handle, with one end flat (not rounded)
- Empty coffee tin with its plastic lid
- 3 cm nails
- Hammer
- Petroleum jelly
- White card
- 2 or 3 bricks or large stones
- Volunteer
- Tin opener

METHOD

1. Hammer the broom handle into the ground so that the flat end is on the top.

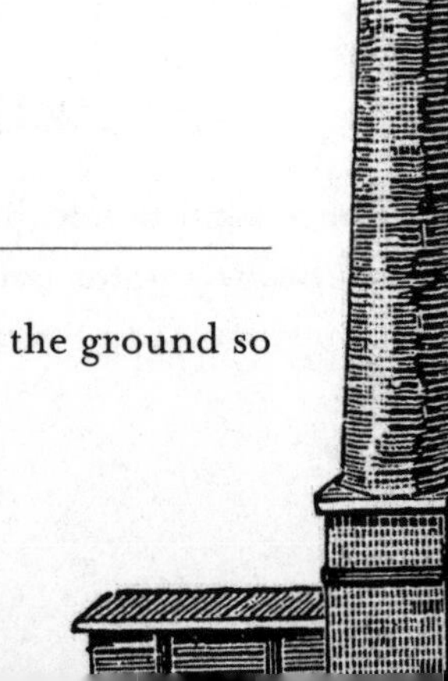

2. Use the tin opener to open the base of the coffee tin.

3. Hammer the plastic lid into the broom handle.

4. Cut out a piece of card just a bit smaller than the coffee tin and place it on the lid.

5. Rub the card liberally with petroleum jelly.

6. Clip the tin onto its lid. (Try to do this upside down, so that the end you just opened meets the plastic lid.)

7. Check to see how secure the arrangement is. If necessary, pound it in a bit more and shore up the base with bricks or stones.

8. Leave for three weeks to a month and then remove the card.

— THE SCIENTIFIC EXCUSE —

Humans send millions of tonnes of waste into the atmosphere each year. Some of this waste is in the form of gases, but much of it is in the form of tiny particles of solids and liquids. Environmental scientists who monitor pollution use special equipment to measure the number of particles falling back down from the sky – you've made a start on a similar study yourself with this experiment.

TAKE CARE!

Try not to position the collector where people might mess around with it. A month is a long time to wait, only to find that someone has dumped a dead frog in your pollution monitor.

Storm in a . . .
BOTTLE

One of the scariest sights on Earth is that of a tornado approaching across level farmland, sucking up and spitting out everything in its path. How would you like to send some shivers down the spines of your family and friends – in fact, anyone who might stumble upon your secret lab? Even if no one's house is likely to be swept away as in *The Wizard of Oz*, you'll have given a hint that your next version could be a little more powerful.

YOU WILL NEED

- Empty 2-litre drinks bottle
- Water
- Sink or basin
- 1 tbsp olive oil

METHOD

1. Fill the bottle almost to the top with water.

2. Add the olive oil.

3. Keep one hand on the opening of the bottle and carefully turn it upside down over the sink or basin.

4. Begin swirling the bottle steadily in a clockwise direction.

5. Take your hand away from the opening but continue to swirl; try to hold the bottle with one hand at the base of the bottle (which is now at the top), and the other close to the mouth but not blocking your view of the water moving inside.

6. The water should be swirling inside, with the darker oil heading down first through the 'twister'.

— THE SCIENTIFIC EXCUSE —

The swirling water inside the bottle is called a vortex, and massive versions of these can form in some severe thunderstorms to create tornadoes. The centre of the vortex is hollow, which allows air to rush up. That air rushing up allows the bottle to empty faster: just think how long it would take to empty the bottle without swirling it. The olive oil is less dense than the water, so it rests on the surface and is the first to be sent spiralling from the vortex.

TAKE CARE!

It's obviously safer to use a plastic bottle than a glass one, but if you are very careful you'll find that a glass bottle is actually better. It's very hard to avoid squeezing a plastic bottle, and that can lessen the tornado effect.

Always Take the Weather

The New Zealand group Crowded House had an international hit in the 1990s singing about taking the weather with them – or was it with you? Could they have been chronicling their scientific efforts through popular music? Even if their song's title is a mere coincidence, you'll have a chance to do some portable weather creation once you've mastered this experiment.

YOU WILL NEED

- One rubber glove
- A wide-rimmed, 3-litre screw-top jar
- Matches
- Water

METHOD

1. Add just enough water to cover the bottom of the jar.

2. Drop a lit match into the jar.

3. Put the glove on, and fit it over the rim of the jar so that the fingers point downward. Keep wearing the glove even after it is sealed around the rim.

4. Pull the glove upwards, but not quite out of the jar: make sure it is still attached around the rim.

5. A foggy cloud will develop instantly.

6. Put the glove back into the jar and watch the cloud disappear.

— THE SCIENTIFIC EXCUSE —

The air inside the jar contains water in the form of water vapour, a gas. When you pull out the glove, the air expands; this expansion also causes the air to cool. And when the air cools, some of the water vapour condenses to create visible particles in the form of the cloud. Pushing the glove back does the opposite, reheating the air and causing the cloud to turn back to invisible vapour.

Why the match? Droplets form more easily around small solids suspended in the air, like the tiny particles of smoke from the match. Farmers who are desperate for rain will sometimes try to 'seed' clouds by spraying small particles of dry ice from planes.

TAKE CARE!

There's no risk involved with this experiment, but take care that you maintain the seal of the glove against the rim to make sure it works according to plan.

MATCH ALERT!

This experiment involves the use of matches and should only be conducted by a responsible adult.

Party Pooper?

It's time now for a bit of sabotage. Try to think of a special event that is coming up and which is likely to have loads of balloons as part of its decoration. A school disco, maybe? A birthday party? Then think of what people would say if they got an unexpected strong smell – say, of garlic:

> But there's no garlic in anything we've got here – just sausages, crisps, cake, jelly. Where is it coming from?

What do you think could be the source of this odour? Read on and be prepared for a surprise.

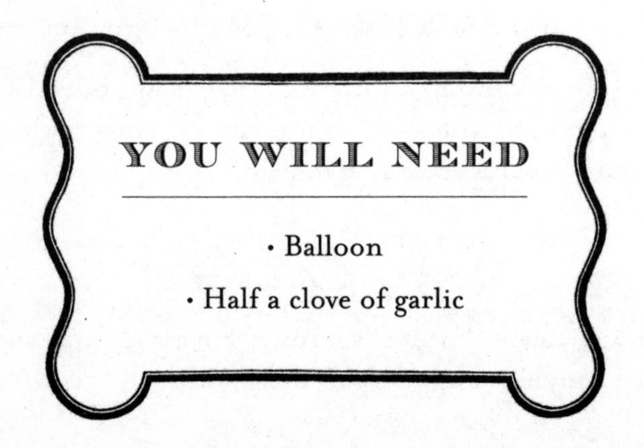

YOU WILL NEED

- Balloon
- Half a clove of garlic

METHOD

1. Wait until most of the balloon decorations are in place and double-check that you have the same sort.

2. Put the garlic in the balloon and inflate it as normal.

3. Tie the end of the balloon and then add it to the decorations, with its knot at the bottom (so the garlic is harder to see).

4. Wash your hands thoroughly so you won't be linked to the mischief.

5. Wait.

6. After a while, a distinct smell of garlic will start to permeate the room.

THE SCIENTIFIC EXCUSE

What we smell is a combination of gases and tiny particles suspended in the air. The garlic provides both, and these find their way through the tiny openings of the balloon when it is fully inflated. (It would take much, much longer for it to seep out through an uninflated balloon because the rubber would not have been stretched to reveal these openings.)

TAKE CARE!

Choose where and when (and to whom) you want to do this – and don't blame the author or publisher of this book. We're only in it for the science – honest!

Well Done!

What's cookin'? Well, for starters there are the nine experiments in the following chapter, which all look to your store cupboard and fridge to get started. When you think about it, just about any cooking involves some sort of chemical reaction: just think about how many ways you can cook an egg. But have you ever thought of how to *fold* an egg? Read on to find out, but make sure you've got all your protective clothing, unlike those poor streakers who found that a takeaway could also involve their getaway car.

FROZEN BUNS

The craze for streaking – dashing naked through a public place – began in the 1970s. Streakers seemed to find their way into all sorts of public places, particularly sporting and entertainment events that were being televised live.

During the 1974 Academy Awards broadcast, for example, Robert Opel streaked across the stage and flashed the peace sign at the millions watching the programme. Presenter David Niven was momentarily stunned but then joked: 'Isn't it fascinating to think that probably the only laugh that man will ever get in his life is by stripping off and showing his shortcomings?'

Most major sports have had their share of streaking attacks as well. Two of the most lasting images of streaking come from Twickenham, the home of English rugby. One of the most reproduced sporting photos in journalistic history shows police escorting 25-year-old Michael

O'Brien from the pitch – with a policeman's helmet strategically placed – during the England-France match in April 1974. Eight years later, Erica Roe disrupted an England-Australia match by streaking topless onto the pitch.

Not all streakers choose to bare all in such public arenas. So it was particularly unlucky that a good-natured streaking trio in Spokane, Washington wound up paying a large price for their prank in January 2004. The three young men (whose names are protected under Washington reporting restrictions) had planned everything to the last detail – or so they thought.

They pulled into the car park of a Denny's restaurant at 5 am. Leaving their car running so that they could make a quick getaway, they jumped out into the –10°C air and dashed through the restaurant wearing only hats and running shoes. But during the brief commotion, one of the Denny's customers left, got into the car and drove it away.

When the police arrived some minutes later, the three young men were huddled together, shivering behind one of the other cars in the car park. 'I don't think they were hiding. I think they were just concealing themselves,' police spokesman Dick Cottam said. 'We always tell people not to leave their car running,' he added.

THE SHY ARSONIST

This little gem of an experiment delights and mystifies within a matter of seconds. No matter how you slice it, the conclusion seems to be that the residue of an experiment is what is needed to get the experiment going in the first place. Which came first – the fire or the ash?

YOU WILL NEED

- A sugar cube
- A brick or other book-sized fireproof object
- Matches
- Ash from a bonfire or fireplace

METHOD

1. Put the sugar cube on the brick (or other fireproof object).

2. Try to light it with a match – it won't catch fire.

3. Spread a little ash over the cube and try again.

4. This time the cube lights easily and will have a lovely blue flame.

The · Scientific · Excuse

The ash in this experiment is obviously vital. Although it remains unchanged after the experiment, it provides enough initial energy for the sugar to combust. Ingredients that 'kick start' experiments, while playing no part in the main chemical reaction, are called catalysts.

· Take Care! ·

Make sure that the fireproof base for this experiment is well away from curtains, furniture or any other household objects that could catch fire.

· Match Alert! ·

This experiment involves the use of matches and should only be conducted by a responsible adult.

– The –
Tipsy Ice Cube

This is an ideal experiment to try out when some guests have arrived and are chatting over a glass of whisky. One look at the madcap ice cube in this glass will have them wondering just how much they've drunk. But be sure to spare them a hangover by explaining the science behind this wicked little trick.

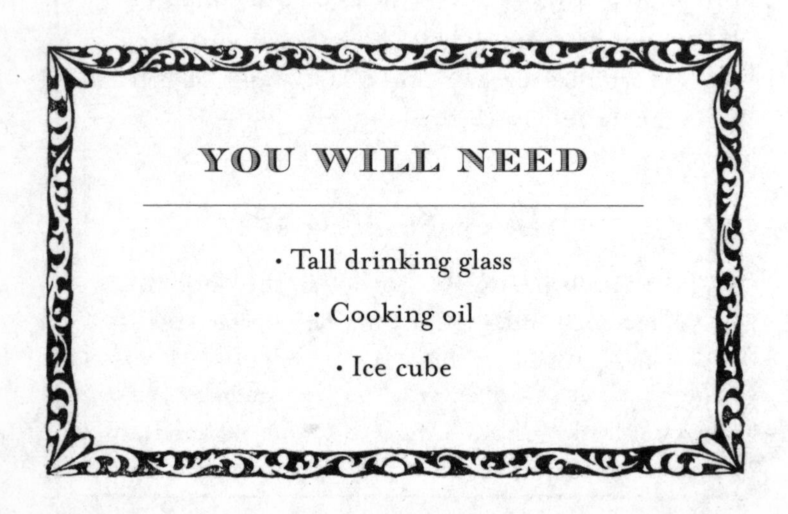

YOU WILL NEED

- Tall drinking glass
- Cooking oil
- Ice cube

METHOD

1. Fill the glass almost to the top with oil.

2. Add an ice cube to the glass.

3. Leave the glass and observe.

4. The ice cube will begin to rock back and forth and even bob up and down without any outside shaking of the glass.

THE SCIENTIFIC EXCUSE

The ice cube and the oil have roughly the same density, so at first the cube sits motionless on the liquid, as it would in a normal drink. But inevitably the ice begins to melt. Water is actually denser than ice (and oil), so a drop of melted water forming at one end of the cube causes the cube to tilt until the drop flows off. If enough water remains on the surface of the ice, the cube will even begin to sink – that is, until the water is washed off and the ice returns to its normal density. Then it floats back to the surface.

TAKE CARE!

Apart from incurring the wrath of a house guest, this experiment is unlikely to get you into any real trouble. Although the experiment itself is easy enough to perfect, you might want to tinker with it a few times beforehand to get a good whisky (or other drink) colour. Try mixing a drop or two of food colouring in with the oil to get the right look.

That's Tasteless

A quick look at this experiment being performed would lead most people to conclude that it was a game of blind man's bluff. But the blindfold is really just an extra precaution. The core of the experiment lies in hoodwinking other senses – taste and smell, to be precise. The results are usually astounding, even if the subject is none too pleased when you bring them up to speed at the end. This is a great experiment in its own right, but so much better if you can get a willing volunteer to be put to the test.

YOU WILL NEED

- Onion (kept in another room until step 4 below)
- Knife
- Blindfold (dark-coloured cloth about 60 cm square, rolled so that it can be tied behind someone's head)
- Handkerchief
- Large clothes peg
- Volunteer
- Witnesses
- Tissue (optional)
- Tweezers

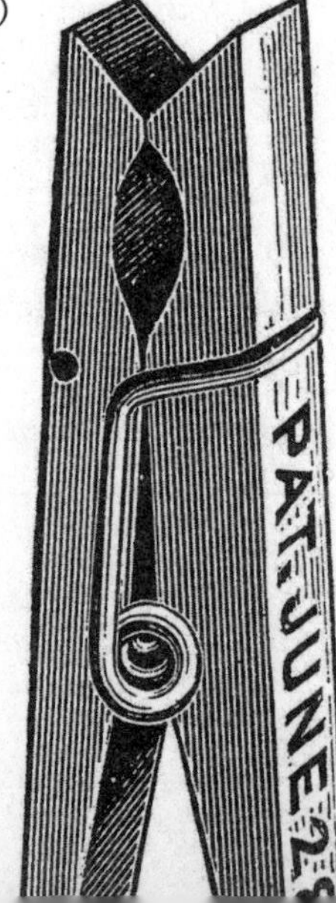

METHOD

1. Reassure your volunteer that everything you do will be safe. Explain that you will need to blindfold them and peg their nostrils shut – but that it won't hurt.

2. Gently tie the blindfold behind your volunteer's head.

3. Take great care pegging the volunteer's nostrils shut with the clothes peg; slide some tissue between the peg and their nose to act as a cushion if it begins to hurt.

4. Go to the other room and cut off a piece of onion about half the size of a stamp. Put it on a small plate.

5. Return to the first room and ask the volunteer to hold out his or her tongue.

6. Use tweezers to place the onion on the middle of the volunteer's tongue.

7. Ask the volunteer to identify the object – they almost certainly won't be able to do so.

8. Unclip the peg on the volunteer's nose.

THE SCIENTIFIC EXCUSE

This experiment demonstrates how we need our sense of smell to identify many tastes accurately – you might have had the same problem when your nose has been stuffed because of a cold.

TAKE CARE!

There's no need to tie the blindfold very tight at all – just make sure it stays over the volunteer's eyes.

The Folding Egg

Recipes often call for the cook to 'fold an egg into the mixture' – usually when people are making some sort of batter or dough. Of course, the word 'fold' has a special meaning in such recipes – to cut and mix lightly to keep air in a mixture. But imagine what fun you could have if someone were reading those words out from a recipe, and you really did fold an egg, right in your hand!

Readers of *Wholly Irresponsible Experiments* might have a head start with this experiment, which shares some of its magic with 'Eggs in the Nude'. But this time it's given a new twist – or fold.

YOU WILL NEED

- Several fresh eggs (only one is needed but you might need a spare or two)
- Drawing pin or sewing pin
- Long needle (more than 6 cm)
- Vinegar
- Bowl
- 400 ml beaker or tall drinking glass

METHOD

1. Make small holes with the pin on either side of the egg.

2. Use the needle to probe the inside of the egg, swirling and mixing the white and yolk together (you want to make it into an easy-to-pour liquid).

3. Blow on one of the holes so that the inside of the egg (the mixed white and yolk) flows into the bowl. You can use this for a real cooking recipe, by the way. The egg should be empty now.

4. Put the empty eggshell carefully into the beaker or glass and cover it with vinegar. You might need to persuade it to sink by holding it down until it fills with liquid.

5. Leave the egg as it is for about a week. Note how bubbles float off its shell.

6. You want all of the outer shell gone (dissolved), so wait until no more bubbles seem to be forming.

7. Take the egg out and rinse it gently under cold water. It is like the 'egg in the nude' except that it is hollow. Gently squeeze the water out and blow it up again as in step 3.

8. At this point, an observer might think that it is a real egg, so when your friend asks you to fold the egg you can hold this one and simply fold it in your hand.

THE SCIENTIFIC EXCUSE

The vinegar, which is a mild acid, eats away at the hard calcium carbonate eggshell to reveal an inner layer – the membrane that acts as a seal around the yolk and white. That membrane is delicate but still strong enough to be held carefully.

TAKE CARE!

Stages 1, 2 and 3 are the hardest in this experiment. Don't be discouraged if it doesn't work the first – or second – time. That's why you have the spare eggs.

Colour Me Confused

It's one thing hoodwinking our sense of taste by denying it the sense of smell (see 'That's tasteless!', page 132). That's a wonderful trick, but you could argue that because those two senses are physically linked, the result wasn't too surprising. Plus, the volunteer in that experiment was blindfolded. How about trying a variation on the theme, only this time letting your subject approach the task with both eyes wide open? Can you still pull a fast one?

— YOU WILL NEED —

- Six identical drinking glasses
- Either six distinctly flavoured but clear fizzy drinks (such as lemonade), or sparkling water with six flavour essences dropped in them (such as mint, almond, coffee etc.)
- Six strongly coloured food colourings
- Volunteer

METHOD

1. Fill each glass with a different drink and line them up. (Note which drink was which so that you can check on the results.)

2. Ask the volunteer to taste each one and try to identify it. Don't give away the right answer, but record the volunteer's.

3. Ask the volunteer to leave the room or turn around.

4. Top up each drink to the same height and then put a few drops of colouring in each one. Make sure the colouring doesn't 'match' the colour we'd normally expect of the drink (make the lemonade blue, for example).

5. Line the drinks up in a different order and ask the volunteer to try again.

6. Check the volunteer's results – they'll probably have got some of them very confused.

THE SCIENTIFIC EXCUSE

Although we don't need to see something in order to taste it, our eyes can send some expected results through to the brain before something even goes into the mouth. Companies recognise this, which is why a lot of those 'E' numbers that appear on lists of ingredients are actually artificial colours to make a food or drink more appealing.

TAKE CARE!

Make sure that the essences that you add are genuine food essences. Also make sure you use them sparingly (only a couple of drops) because the flavour is so concentrated.

– The – DNA Detective

'A half-eaten banana was found near the door of the safe that had been blown open.' What would Sherlock Holmes have done with that piece of evidence, found at the scene of a multi-million-pound bank robbery? Even he would have been stumped if there were no way of linking the banana with the suspect being held in the police cells.

Modern detectives have a few more tools to use, and one of them is DNA analysis. DNA is the chemical 'blueprint' of every living thing – what makes it different from every other living thing. People leave traces of their DNA on things that they touch or eat, so the first step is to extract the DNA from the banana that the safe-cracker was eating. But how do you set about doing that? Elementary, my dear Watson.

YOU WILL NEED

- Half-eaten banana
- Food processor
- Salt
- Cup
- Hot water
- Washing-up liquid
- Methylated spirits
- Paper coffee filter
- Wooden skewer
- Clear drinking glass
- Teaspoon

METHOD

1. Put the banana into the food processor.
2. Add a teaspoon of salt to half a cup of hot water.
3. Pour the salt-water mixture in with the banana and process for one minute. Rinse the spoon and cup.
4. Pour the mixture through a sieve back into the cup. Press the back of the spoon against the sieve to push the solids through.
5. Add a teaspoon of washing-up liquid. Stir occasionally for 5 minutes.
6. Set the coffee filter over the drinking glass and pour the mixture into it, so that it drips into the glass.
7. Slowly pour methylated spirits down the side of the glass until it forms a 1-cm layer on the top of the mixture.
8. A layer of fluffy white material should form between the mixture and the methylated spirits.
9. Use the toothpick to extract some of this material (which is the DNA); with any luck, you'll be able to pick out a strand of DNA.

THE SCIENTIFIC EXCUSE

DNA, like other chemicals, is stored inside cells. Processing the banana helped break down some of these protective barriers; the washing-up liquid helped dissolve some other chemicals that were locking the DNA inside. DNA dissolves inside water-based mixtures, but doesn't dissolve in alcohol. The fluffy material at the end of the experiment was DNA that was not dissolving in the alcohol.

TAKE CARE!

Make sure that it is an adult who handles and pours the methylated spirits.

Non-Stick Pan?

Just in case you're performing the experiments from this book in order, you'll probably need a little light relief after all your forensic efforts in the DNA experiment (see previous pages). How about something that's funny, informative, challenging ... and quick? It might also let you get your own back on that muscle-bound friend who's always bragging about how strong he is.

If you're planning to do this one as a trick, finish steps 1–3 before calling your friend in.

YOU WILL NEED

- Perfectly smooth work surface (formica or vinyl, or even marble will work)
- High-sided baking tray
- Water
- Cloth

METHOD

1. Pour some water (equivalent to about half the area of the baking pan) onto the work surface.

2. Spread this out to match the shape of the base of the baking tray. Add a little more water to make sure all the whole of the underside of the tray will be wet if you put it down there.

3. Put the baking tray down on that wet spot, and dry any bits that jut out from under the pan.

4. Pretend to be reading a recipe and ask your friend to pick up the baking tray carefully, to see how much it weighs (let them puzzle over why you'd need to know that).

5. Watch as their face goes through a bizarre set of expressions ... the baking tray seems glued to the counter!

THE SCIENTIFIC EXCUSE

A scientific principle called adhesion causes water to seem sticky (don't forget, the real name for sticky tape is 'adhesive tape'). This stickiness can be up to 200 times stronger than normal air pressure, which we know is pretty strong because it powers a lot of the experiments in this book.

TAKE CARE!

Don't provoke anyone who's too strong – in case they really take it the wrong way.

PEPPER POINTERS

From time to time, scientists need to show others who's boss. It's only human nature. Here's a chance to demonstrate to your family and friends that you don't really need to play by all the rules – putting dirty clothes in the laundry basket, making your bed, that sort of thing. Why? Because you can simply issue commands and household objects obey your every word.

YOU WILL NEED

- Small mixing bowl or cereal bowl
- Black pepper
- Water
- Washing-up liquid

METHOD

1. Fill the bowl almost to the top with water.

2. Sprinkle some pepper on the surface, mainly near the centre.

3. Rub a little washing-up liquid on your fingertip (hide this movement if you're doing this as a trick).

4. Shout 'Be off!' or 'Away with you!' at the pepper, and dip your finger in the middle of the bowl.

5. The pepper will rush away from your finger and over to the side of the bowl.

— THE SCIENTIFIC EXCUSE —

Water molecules are attracted to each other very strongly, but even a small amount of soap can break those bonds. The soap chases through the water molecules, getting in between them. However, as the water molecules nearest to the edge are not yet affected by the soap, they are still strong enough to tug the pepper-holding water molecules to the side.

— TAKE CARE! —

Make sure that you keep your supply of washing-up liquid hidden if you're planning to do this as a trick. Just leave a coin-sized drop on the counter and dip your hand in it just after step 2 (and while no one is looking).

GLOWING GRAPES

Pressed for time and keen to show off some of your powers as a wizard? Try this spectacular experiment, which will produce glowing results in just a few seconds. The materials couldn't be simpler, but you might want to bone up on your electromagnetism if you want to provide a satisfying explanation for this bit of kitchen 'lightning'.

YOU WILL NEED

- Microwave oven
- Seedless grapes
- Sharp knife
- Microwave-safe plate or saucer

METHOD

1. Cut a single grape almost into two halves, leaving a small flap of skin connecting the halves.
2. Put the connected grape halves on a microwave-safe plate or saucer and place in the microwave oven.
3. If the oven has a turntable, try to position the grapes in the exact centre.
4. Turn the oven on at high power for ten seconds.
5. Within about five seconds, a bright light (accompanied by an audible buzzing) develops in the 'bridge' between the grape halves.

6. After a few more seconds, the grapes begin to emit sparks and a glowing arc or cloud rises from them. The effects then disappear.

— THE SCIENTIFIC EXCUSE —

Although this experiment is one of the simplest in this book to perform, the scientific explanation is probably the longest and most complicated. Boiled down (excuse the pun), the science is as follows:

First of all, note that grapes are reasonable conductors of electricity. That means that when the microwave is turned on, it not only heats the water inside the grapes but also causes some electrical charge to move from grape half to grape half across the skin 'bridge'. This electrical current releases energy as it passes through, so the bridge becomes very hot, eventually catching fire.

As the electrical current now passes through this flame, it ionises (removes electrons from) the gas. At this point, the gas (flame) is conducting electricity; it produces a glowing arc or cloud, a similar effect to the St Elmo's Fire that sailors observe on ships' masts in stormy weather.

The effect dies down when all of the water from the grapes has evaporated, killing the flow of charge, or when the power timer shuts the oven off.

TAKE CARE!

Be warned: there is a slight chance that this experiment will damage the microwave. As a basic precaution, don't be tempted to leave the microwave on for longer than the ten seconds.

Generally Speaking

Most people tend to study one scientific subject at a time when they are at school. Chemistry labs are full of Bunsen burners, test tubes, strange liquid mixtures bubbling away and students trying to decide what colour a piece of paper has turned. Biology and physics departments have their own look as well, what with diagrams of plant cells or inclined planes and stop-watches. But when 'things happen' in real life, you can't always pigeon-hole them. Nor can you package and store these experiments on a particular shelf, unless it's the one marked 'mayhem'.

BRIDGE OF SMOOTS

The Harvard Bridge spans the Charles River in Massachusetts where the river is at its widest, forming a basin that is a haven for sailing boats and rowers. On one side of the river is Boston, its old-and-new skyline forming a charming backdrop. On the other is the separate city of Cambridge, home of two of the world's great universities: Harvard and Massachusetts Institute of Technology (MIT).

The low-lying bridge offers excellent views of both cities, as well as the most direct way home for students after a night out in Boston. The only drawback is that it is long, and the late-night trudge home – often in the teeth of a howling blizzard – can seem endless. If only there were some way that walkers could keep track of their progress.

That aim – to provide evenly spaced markers painted on the bridge's pavement – prompted some MIT students to invent a new measurement in 1958. Oliver Smoot, a first-year student, was chosen to measure the bridge. More accurately, he was used to measure the bridge. Fellow students took the 1.7 m-tall Smoot and laid him on the pavement, moving him head over heels and then extending him again and again, all the way across the bridge. Smoot was chosen because his name sounded 'scientific', like 'watt' or 'amp'. Student volunteers painted coloured lines to mark every ten 'smoots'.

All of this would have been ancient history if the Massachusetts Department of Public Works had had its way in 1987, replacing the concrete with unpainted new slabs. There was an outcry in the local press, echoed by the Cambridge police (who used smoots as references for accidents on the bridge). In the end, the concrete was replaced, but in 1-smoot rather than the normal 6-foot (1.83 m) lengths.

MIT students continue to repaint the 10-smoot markers right to the Cambridge end, which is marked '364.4 smoots and one ear'. And what happened to Smoot himself? Well, in 2005 he retired from the board of the American National Standards Institute, which sets standard units for American industries.

Mad Molecules

This delightfully simple experiment can be used as a challenge or bet for a young audience. But before you smile at these confused youngsters with a superior air, you should remember that experiments such as this one helped open up our modern understanding of science. Since the mid-1600s, some of the greatest names in science – including Robert Boyle, Antoine Lavoisier and John Dalton – have devoted themselves to this area of science. You can simply reap the results of their labours, and win the occasional bet as a result.

You Will Need

- 2 small mixing bowls
- Hot water
- Cold water
- Food colouring
- Young audience

METHOD

1. Fill one bowl with hot water and the other with cold.

2. Put the bowls side by side.

3. Ask your audience to predict what will happen if you add a drop of food colouring to each bowl.

4. Challenge anyone who says 'it will look the same for both' to a bet.

5. Add the drop of food colouring to each bowl.

6. Note how the food colouring in the hot water fans out rapidly, while the one in the cold stays pretty much the same.

7. Collect your winnings.

THE SCIENTIFIC EXCUSE

As substances warm up, the molecules making them up do so become more excited (they move around more). That means that the hot water molecules are more active than the cold ones, which is why they take the colouring for a ride around the bowl. But it was looking in the other direction – to see whether there is a point where things are so cold that nothing moves – that occupied those great scientists. And indeed there is: absolute zero (minus 273 degrees Celsius, or minus 459 degrees Fahrenheit).

TAKE CARE!

Make sure none of the younger children dips a hand in the hot water.

Paper Saucepan

Who in their right mind would set about cooking with paper over an open flame? An irresponsible cook, obviously. Just ask any of your friends. Ask them another question, though: would it still be *possible* to boil water in a paper cup over an open flame? Here's a demonstration of how you can silence the doubters.

YOU WILL NEED

- Paper cup (must be unwaxed)
- Eight red builder's bricks
- 60-cm string
- Short candle
- Water
- Fork or knife (to poke holes in the cup)
- Matches or lighter

METHOD

1. Poke two small holes opposite each other near the top of the cup.
2. Thread the string through the cup so that there is a similar amount on each side of it.
3. Make two piles of four bricks, about 40 cm apart.
4. Suspend the cup between the brick piles, anchoring it in

place by sandwiching the string between the third and fourth courses of bricks.

5. Knot the string on the outside of each brick tower, so that the string remains taut.

6. Position the candle below the cup with about a 5cm gap between the wick and the base of the cup.

7. Fill the cup about three-quarters up with water.

8. Light the candle.

9. The water will heat up and eventually boil – but the paper cup won't catch fire!

— THE SCIENTIFIC EXCUSE —

For paper – or anything – to burn, it must warm up to its auto-ignition temperature, or 'kindling point'. Put simply, that is the lowest temperature at which a solid will spontaneously combust. A candle's flame would normally be enough to reach that temperature for paper, but the water draws heat away from the paper, keeping the paper below its kindling point while the water itself heats up to boiling point.

TAKE CARE!

This experiment is all about heat passing through the paper to the water that will absorb it, so you must not use a wax-covered paper cup. Why? Because the wax will absorb the heat as much as the water does, but in a much smaller (and therefore easier to heat up) volume ... increasing the chances of the paper itself catching fire.

MATCH ALERT!

This experiment involves the use of matches and should only be conducted by a responsible adult.

– THE – DISAPPEARING BEAKER

This experiment is so – dare we say it? – reliable and easy to perform that it demands an audience. Who could possibly misplace a graduated beaker inside another one? But there's no getting around the fact that the smaller one really did go inside the larger one, and no one took it out. And it's nowhere to be seen. Hmm.

- 400 ml cylindrical glass beaker
- 250 ml cylindrical glass beaker
- Bottle of glycerine (sold as a throat soother)

METHOD

1. Put the smaller beaker inside the larger one.

2. Show the beakers to your audience (the smaller beaker should be easily visible).

3. Fill the gap between the beakers with glycerine.

4. The smaller beaker will have disappeared!

THE SCIENTIFIC EXCUSE

This experiment is all about refraction, the process by which light waves bend as they pass through different substances. Refraction can be measured, and the result is called the refractive index of the substance. The glycerine and the glass have very similar refractive indices, which means that light passing through the glass continues straight through the glycerine. The same trick would not work with water between the beakers, because its refractive index is different from that of glass – making the other beaker clearly visible. Likewise with air, as you saw before you added the glycerine.

TAKE CARE!

Make sure you use glass beakers of the cylindrical shape, not the cone-shaped beakers with the narrow tops.

Dracula's Cocktail

'Just one draught of this lovely liquid and I feel I could sleep for eternity.' You might imagine Dracula himself uttering those words if you master this do-it-yourself blood exploit. It's fun trying to match your kitchen ingredients to a vampire's menu, but even more fun if you take things further and actually drink from the goblet. Ahh, now time for just one little bite ...

- Mixing bowl
- 150 ml corn syrup
- 50 ml water
- 5 tbsp cornflour
- 4–5 tsp red food colouring
- 2–3 drops of green food colouring
- Wine glass (optional)

METHOD

1. Mix the cornflour thoroughly with the water.

2. Add the corn syrup and stir well.

3. Stir in 3 tsp of red food colouring.

4. Add 2 drops of green food colouring.

5. Check on the colour of the 'blood': if it is too light, add one or two more tsp of red colouring. Add another drop of green if it is still a little pink.

6. The 'blood' is ready. You can pour some into the wine glass and have a Dracula cocktail.

THE SCIENTIFIC EXCUSE

There's no mysterious science at work here; only household ingredients mixed to get the right colour and consistency of the real thing. Bear in mind that real blood is a little darker (browner) than you might think, so don't go for pure red.

TAKE CARE!

You'll have enough explaining to do anyway ('why is there blood all over the place?'), so don't add to your worries by getting any clothes or furniture stained.

– The – Unbalanced Balance

On balance, this is one of the most eye-catching experiments that you will ever perform. What's more, it cries out for an audience. And like a champion boxer, you'll be packing a one-two punch. The first is getting the 'mating forks' to balance on a toothpick. But the real knockout is burning away that toothpick to nothing ... and keeping the forks balanced in mid-air!

YOU WILL NEED

- 2 forks, or a fork and a spoon
- Toothpick
- Heavy-bottomed drinking glass

METHOD

1. Link the forks together by weaving their tines (this takes a little practice and might be easier if you use a fork and a spoon).

2. The handles should be pointing away from each other, not quite forming a straight line (more like a broad 'V').

3. Work the toothpick into the junction of the two forks until about 5–10 mm emerges from the back of the forks.

4. Balance the forks/toothpick combination on the rim of the glass. This also takes a little practice but it can be done. The point of the fork 'V' is suspended outside the glass with the two arms extending inward. About half of the toothpick juts into the inside of the glass.

5. Light the tip of the toothpick that is inside the glass.

6. The toothpick will burn away until it reaches the rim.

7. No part of the toothpick extends in over the glass anymore, but the forks will remain balanced!

— THE SCIENTIFIC EXCUSE —

This outlandish gravity-defying display depends on the notion of centre of gravity. Think of how a tightrope walker remains balanced so high above the circus Big Top. The chances are that he makes his way along the wire while holding a long pole. The fork combination itself, rather than the toothpick, acts as the tightrope walker's pole – providing balance in the unlikeliest circumstances. Even when the toothpick has largely burned away, there is still a very small point on which the arrangement rests. Why does the burning stop when it reaches the inside rim? Probably because the glass absorbs the heat and lowers the temperature.

MATCH ALERT!

This experiment involves the use of matches and should only be conducted by a responsible adult.

The Finger of Fate

Are you looking for another one of those quick 'how in the world?' demonstrations? A quickie that demonstrates your supernatural powers – or maybe your superhuman strength? Now read on. And while we're on the subjects of commands: don't get up until I say so. Is that clear?

YOU WILL NEED

- Willing volunteer
- Upright chair

METHOD

1. Have your friend sit in the chair with good posture – head back and chin up.

2. Put your index finger up to his forehead and press lightly but firmly.

3. Ask your friend to get up.

4. He won't be able to stand, or even move very much.

— THE SCIENTIFIC EXCUSE —

You might well have supernatural powers or superhuman strength, but you won't need either of them for this experiment. It's all down to centre of gravity, or, more accurately, centre of mass. That is the point at which the mass of an object seems to be concentrated (it's also part of the secret of tightrope walkers). The centre of gravity of a seated person is the chair. In order to stand, the person must move that centre of gravity over the feet. And the first step for that to happen is to move the head forward. But even the slight force of your fingertip is enough to keep the cards stacked against the person moving out and up. Your friend stays put.

TAKE CARE!

There's no danger here – just the usual warning about not overplaying your hand and getting on your victim's nerves.

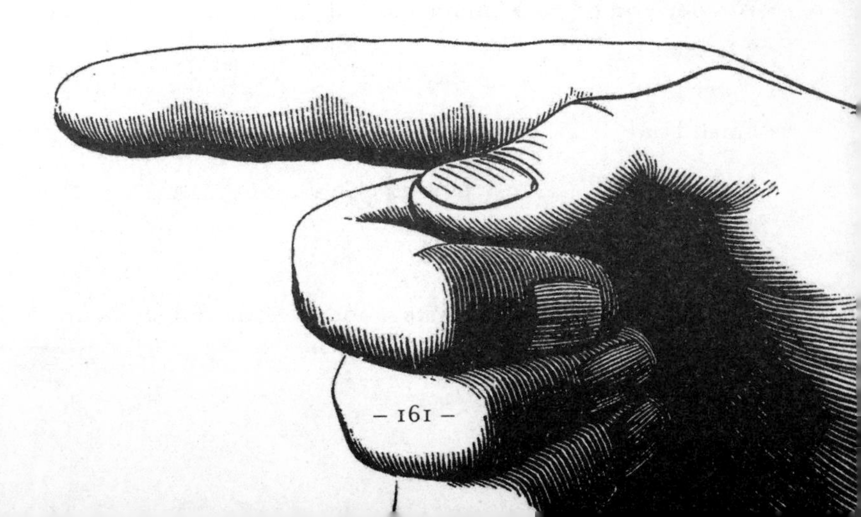

ANTI-FLAME THREAD

Scientists often get the best results when they seem to be confounding nature – lids not falling off when they should, foam bubbling up out of nowhere, pepper obeying their commands. This experiment falls into that category, and you'll soon see why! Bear in mind, though, that it calls for extremely calm conditions, so it is best done indoors. If you do plan to do it outside, make sure you have chosen a day with no wind and that you're shielded from even the slightest breeze.

YOU WILL NEED

- Cotton thread
- Wine cork
- Knife
- Matches
- Wooden rod or pole (about 1 m long)
- 3 tsp salt
- Water
- Small bowl

METHOD

1. Cut about 40 cm of thread.

2. Half-fill the bowl with water, add the salt and stir it in until it dissolves.

3. Soak the thread in the salt solution and then take it out to dry.

4. Repeat step 3 about three more times.

5. Cut notches in the cork and tie one end of the thread to it.

6. Rest the long rod or pole across two chairs or between a chair and a counter.

7. Tie the other end of the thread to the rod. The cork should be hanging down from it.

8. Light the thread at the lower end, near the cork.

9. The thread will burn away, leaving a thin trail of ash. But this ash is still holding the cork up!

— THE SCIENTIFIC EXCUSE —

You alone know that although the cotton thread has burned away, the cork is still held up by a column of salt that runs the length of the thread. It is strong enough to withstand the pull of the cork, but a slight breeze would break this magic chain.

TAKE CARE!

Make sure you don't perform this experiment near to anything that could be damaged by flames.

MATCH ALERT!

This experiment involves the use of matches and should only be conducted by a responsible adult.

RULE(R)S ARE MADE TO BE BROKEN

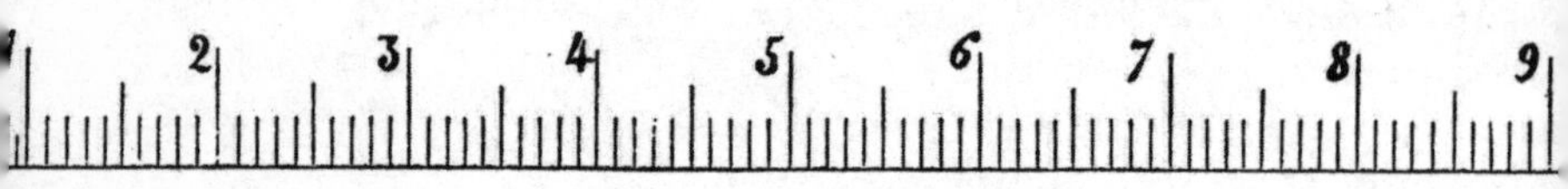

Students in some schools traditionally break their rulers at the end of their last day at school. In no way does this book encourage such wilfully destructive behaviour. It does, however, encourage the spread of knowledge and scientific curiosity. That means it's all right to break not just one ruler, but maybe even two.

YOU WILL NEED

- 2 rulers, preferably wooden (which will be broken)
- Empty golden syrup tin (with its metal lid)
- Strong tape
- Blu-Tack
- Metal bottle cap (from a drink needing a bottle opener)
- Volunteer
- Water

METHOD

1. Fill the tin to overflowing with water and fix its lid back on tightly.

2. Fix the bottle cap to the top of the lid (right-side up) with a little Blu-Tack.

3. Put one ruler on the counter and centre the tin on it.

4. Rest the second ruler across the top of the tin (directly above the first ruler) and secure it with more Blu-Tack.

5. Have a volunteer hold this combination up and loop the tape tightly around the rulers. You should have a loop at each end, tightly pulling the rulers towards each other.

6. Put this set-up in the freezer and wait 24 hours.

7. Open the freezer and note how the top ruler has been broken; the bottom one might also be broken.

THE SCIENTIFIC EXCUSE

Most chemical substances contract (become smaller) as they cool. Water is no exception, until it starts to get close to freezing point (0 °C). That's when it seems to march to the beat of a different drum and begins to expand. By the time it is a solid (ice) it takes up more volume – enough to push the lid out until it breaks free of the ruler binding it.

TAKE CARE!

It should be obvious, but make sure that the rulers you use in this experiment are your own, and that you can live without them.

Skim Reading?

Wouldn't you love to be able to walk into a room, pick up the first couple of things that came to hand and then demonstrate some astounding scientific principle? This experiment – or demonstration – falls into that category. It's what you might call a slow-burner: why not keep it in mind so that next time you walk past a friend's bookshelf, you can dazzle them.

YOU WILL NEED

- 2 paperback books of about the same size and number of pages

METHOD

1. You're aiming to 'lock' the books together by having their pages overlap each other.

2. Put the books on a table, facing each other so that their open-pages sides just touch.

3. Lift each book up by the open-pages side, so that the spines stay on the table but edge closer to each other by about 3 cm.

4. Rifle through the pages with your thumbs (from the back of the book to the front).

5. If you've managed to do this right, the pages of the books will overlap each other by the same 3 cm.

6. Try to pull the books apart. It seems as though they're locked together.

THE SCIENTIFIC EXCUSE

This experiment works thanks to friction, the force that acts as a brake against things sliding apart. Each overlap is a source of friction, but multiplying that force by 50, 60 or however many overlaps you manage to produce will increase the force a great deal. No sweat.

TAKE CARE!

No real problems here – except don't use someone's priceless first edition (even if it is a paperback).

Zero-G Cola

It's really best to do this trick outdoors, for two reasons. First, you'll find yourself pouring out a cupful of water in the first section, and you don't want to get that all over the place. Second, the 'tricky' part of this experiment – catching it while it falls – will probably take a bit of practice.

YOU WILL NEED

- Polystyrene cup (as used for hot drinks)
- Pencil
- Water
- Cola
- Audience

METHOD

1. Fill the cup with water and stand it where it can be easily seen.

2. Poke a hole in the side of the cup, near the bottom, with the pencil.

3. Water will gush out, emptying the cup in a couple of seconds.

4. Now tell your audience that you will defy gravity by keeping most of some cola in the cup even though it will have the same hole.

5. Hold your finger over the hole while a friend fills the cup with cola.

6. Keeping the hole (still blocked with your finger) towards your audience, hold the cup out at arm's length.

7. Note very carefully exactly which way the hole is pointing, and after a short countdown let go of the cup.

8. Here's the tough part: no cola will spill as the cup falls, but it's up to you to catch it (and plug the hole again) before it hits the ground.

9. If you do it right, you can show your audience the safely plugged cup with all of the cola still in it.

— THE SCIENTIFIC EXCUSE —

The cola – or anything inside the cup – becomes 'weightless' while it falls and therefore doesn't spill through the hole. Of course it's not really weightless – it's simply falling at the same speed as the cup. The water 'falls' through the hole of the motionless cup at the same speed, but the cup doesn't keep pace.

TAKE CARE!

Make sure you practise the second phase of this experiment – the 'catching it while it falls' payoff – with water several times before you try it with cola. It calls for a special knack of following the cup with your hand and then darting in and grabbing it.

Edible Crystals

Every science book needs to be able to lighten some of the theoretical content one way or another. It could be with groan-inducing puns (we plead guilty), outlandish or unexpected results (guilty again, your honour), or letting you eat the fruits of your labours. And that's where this excellent introduction to crystals comes in. Come and crunch away on some sweet science.

YOU WILL NEED

- 150 g sugar
- 200 ml water
- Saucepan
- Wooden spoon
- Heatproof bowl
- Pencil
- 40 cm length of string

METHOD

1. Bring the water to the boil in the saucepan.

2. Add the sugar and stir while the sugar dissolves.

3. Pour this mixture into the bowl.

4. Tie one end of the string to the pencil and let the other end dangle in the sugar-water solution.

5. Keep the pencil supported with books on either side while the string continues to hang down into the bowl.

6. Within a few days crystals will begin to form on the string.

— THE SCIENTIFIC EXCUSE —

The science is simple enough in this easy-to-manage experiment. It's all about solutions. The sugar dissolves (goes into solution) thanks to the heat and with a little help from your stirring. Then the liquid travels up the string through the process called osmosis – the same way plants suck water from the soil. Then the water evaporates, leaving the solid sugar crystals – no longer in solution – there to be plucked from the string.

— TAKE CARE! —

Be careful whenever dealing with boiling water. Only an adult should handle that section of the experiment.

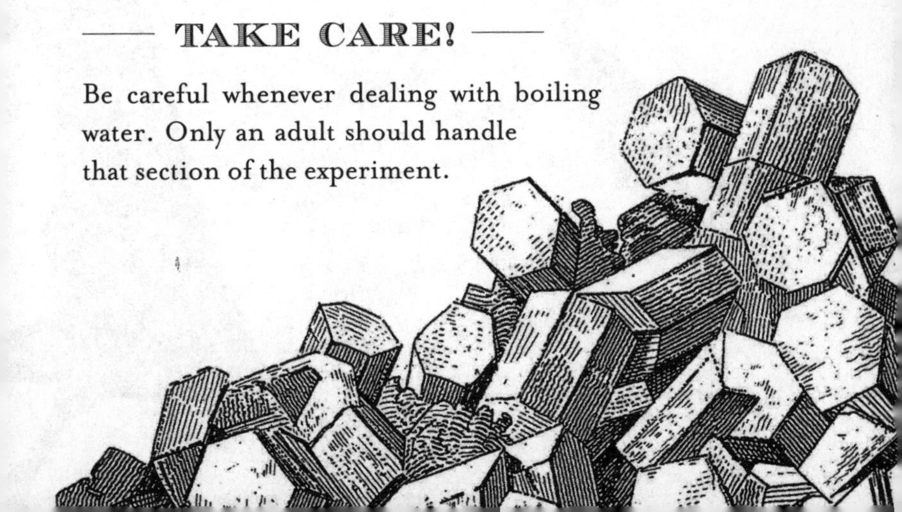

Saloon Brawl Prop

Wouldn't it be wonderful to have some of the props that Hollywood directors have at their disposal? Not CGI effects, which soon seem artificial. More like something that you could use to hit a real person over the head, and then see shatter (the prop, not the head). This demonstration helps you do just that – you'll cook up a batch of realistic glass or ceramic tiling. Whatever cinematic role the end result plays, it will smash to smithereens when it is slammed into someone's head. Lights, camera, action!

METHOD

1. Use butter to grease the base of the baking sheet.

2. Put the baking sheet in the refrigerator.

3. Pour the sugar into the saucepan and put it on the stove to heat gently.

4. Stir the sugar frequently as it warms and cooks.

5. When the sugar has melted, and before it begins to change colour, get the baking sheet out and pour the melted sugar onto it.

6. Shape the still-liquid sugar into squares and rectangles with the spatula or slice.

7. Leave the tray to cool on a counter.

8. In about an hour, it will have hardened and become brittle – ideal for filming that brawl at the local saloon.

THE SCIENTIFIC EXCUSE

There's no surprising science at work in this experiment, just sugar molecules melting as they are heated and then becoming solid once more when they cool down again.

TAKE CARE!

There's no real danger arising from this deceptive trick, but never use the 'glass' to hit anyone on the face – just the back of the head, as in the movies.

Cliffhanger Ending

Here's an experiment that you could do in the classroom, ideally while waiting for your English teacher to arrive. You'll find that it works best with a lot of identical books, and what better place to find identical books than in an English lesson, if you're all reading the same set text? And if your teacher arrives in the middle of this demonstration? Simple – just tell her that you are seeing whether or not the book really is a cliffhanger.

YOU WILL NEED

- 6 identical books (not too thick)
- Level desk, counter or table

METHOD

1. Arrange the books in a neat pile at the edge of the desk, with the narrow sides of the books along the edge.

2. Slide the top book out slowly until it teeters on the pile, then slide it back a little.

3. Move the second book out (with the top book still on it) until they both teeter, then slide them back a bit.

4. Continue this until you have moved every book.

5. When you finish, the top book will appear to be suspended in mid-air, like a cartoon character who has just run off a cliff and is about to fall.

THE SCIENTIFIC EXCUSE

You've just demonstrated the centre of gravity (or centre of mass), the 'concentration point' of the overall weight, where the whole arrangement balances. Although the top book seems as though it should be about to plummet, more than half the weight of the books supporting it is safely on the desk.

TAKE CARE!

Just make sure you choose the right books, and the right teacher. A good sense of humour is a help (for the teacher, that is).

Create a Fossil

Every so often, a TV chef will pull out a lovely roast chicken or a bubbling curry from off-camera and say 'Here's one I started earlier'. It makes sense: after all, you can't watch the television for the entire six hours it takes to cook a turkey. But how would you feel if you could say 'Here's one I started about 40 million years ago'? That might make your audience stop in its tracks. And you'd only be stretching the truth a little bit. You've been recreating all the stages involved in turning something into a fossil and just given them a little nudge.

YOU WILL NEED

- Pack of bath salts
- Warm water
- Small baking tray
- Bath sponge
- Fine sand (sandpit sand works well)
- Cup
- Empty plastic margarine tub
- Knife
- Scissors

METHOD

1. Use the scissors to cut the sponge into a three-dimensional shape that will become the fossil: it could be in the shape of a dinosaur bone or tooth, or maybe even a primitive shellfish. Make sure the finished product can fit easily into the tub.

2. Make a few small holes in the bottom of the tub and rest it on the tray.

3. Fill the tub with a 1–2-cm layer of sand, and then lay your sponge 'fossil' on it.

4. Cover this with another 3-cm layer of sand.

5. Mix 4 tbsp of warm water with 4 tbsp of bath salts in the cup; pour this mixture over the sandy tub, letting the liquid sink through the sand.

6. Leave the tub in a warm place, such as the bottom of the airing cupboard or a sunny windowsill, for five days.

7. Repeat step 5 for each of those days.

8. Leave the tub without adding any more liquid for another two days.

9. Carefully remove your fossil, which should be very hard by now.

— THE SCIENTIFIC EXCUSE —

Sponge, as we all know, absorbs a great deal of water. The salty liquid seeps through the sand, being collected by the sponge on its way down and then being sucked back up from the tray by the process called osmosis. Then it evaporates, only to be topped up every day by you. Each time the water evaporates, it leaves some of the salts that were in solution with the water. Each 'dose of salts' hardens the fossil a bit more, until it really does look 40 million years old.

TAKE CARE!

If your fossil still feels a little pliable, leave it to dry for another day.

Pharaoh's Secret

If you found that your house contained some fossils dating back millions of years (see 'Create a fossil', page 176), why shouldn't it contain some equally mysterious messages from only about 4,000 years ago – from ancient Egypt, perhaps? How is it that no one knew about these secrets? Who can unravel them? Only you have the answers, and the skills to penetrate these mysteries.

YOU WILL NEED

- Half a lemon
- Egg cup
- Toothpick or fountain pen
- Candle
- Piece of ordinary writing paper (the older-looking the better)

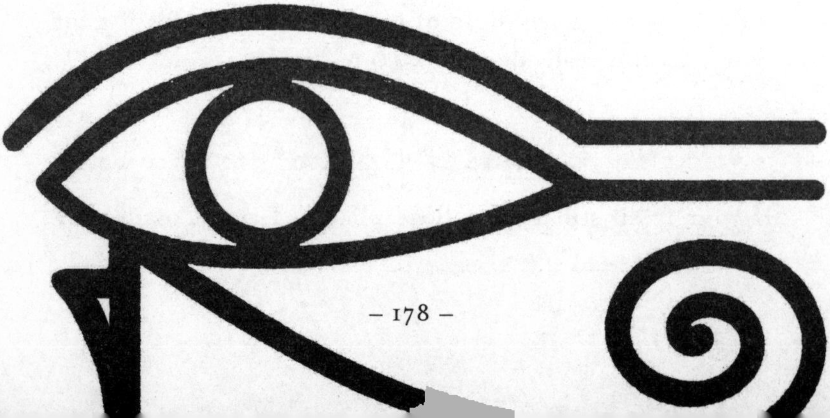

METHOD

1. Squeeze the juice of the lemon half into the egg cup.

2. Write a message using ancient Egyptian symbols, known as hieroglyphs, on the piece of paper.

3. Pass the paper to your archaeologist partner, and see if he or she can read it. The lemon juice will have evaporated and the paper will look untouched.

4. Light a candle and hold the back of the paper close to the flame, but not too close.

5. The ancient message – the Pharaoh's secret – will be revealed as a brown ink message.

THE SCIENTIFIC EXCUSE

The evaporating lemon juice that left no visible trace is just part of the secret. But it leaves behind citric acid, which has a lower ignition temperature (it gets ready to burn, turning brown) than the neighbouring paper. That is why the message 'burns' its way through.

TAKE CARE!

Don't rush and get the paper too close to the flame; be patient and edge it closer until the message appears.

MATCH ALERT!

This experiment involves the use of matches and should only be conducted by a responsible adult.

Cash and Carry

'He demanded the ransom in cold, hard cash.' 'That money seems to be burning a hole through your pocket.'

The English language is full of expressions linking money with temperature. Why is that? And why do these expressions contradict each other? Is there any way of performing a test on some money to get to the root of all this?

This experiment should shed some light on why we think of money in the way we do. The results might surprise you.

YOU WILL NEED

- Loose blindfold (a large handkerchief will work)
- 2 identical coins (British 10 p coins work well)
- Ice cubes or a freezer
- Towel
- Volunteer

METHOD

1. Blindfold your volunteer and tell him that he is going to be a human set of scales.

2. Have him put his arms out with palms up and middle and index fingers outstretched.

3. Put a coin on each pair of outstretched fingertips and ask him to decide which weighs more. He should say that they weigh the same.

4. Put one of the coins on an ice cube or in the freezer for two minutes and then dry it quickly with the towel. Don't tell your volunteer what you are doing.

5. Tell your 'scales' that you have two other coins to weigh (they are, in fact, the same two coins).

6. See which of this pair seems heavier. He will probably say that the colder one does.

THE SCIENTIFIC EXCUSE

This simple experiment actually has a mystery at its heart. Scientists aren't exactly sure why colder objects feel heavier than warm ones. One theory is that the nerve endings that detect changes in weight are also used to detect changes in temperature and that the result is something of a 'short circuit'.

TAKE CARE!

This is in competition for the least risky experiment in the book.

Scrambled Bed

You've had a run of wet weather and your mum has only now managed to get the sheets dry on the line outside. What can you do to help? How about showing off a bit of physics by making one of the sheets a target and hurling an egg at it? It's perfectly safe, of course. Admittedly, you might lose your nerve once you get the clean sheet unfurled, worried that you might wind up with egg on your face. Read below for a safe way to practise.

YOU WILL NEED

- Egg
- Bed sheet
- Golf ball (optional)
- Two volunteers

METHOD

1. Have your volunteers hold the sheet so that there is enough slack to make a trough.

2. Pick up the egg and take aim at the trough of the sheet.

3. Throw the egg as hard as you can.

4. The egg won't break.

— THE SCIENTIFIC EXCUSE —

You could produce a wonderfully long and complicated equation to explain this experiment, but the basic principle is simple. If you increase the amount of time needed to stop a moving object, you need less force to do so. Even the extra tenth of a second or so that you give the egg – compared to hitting an unyielding solid object – is enough to let the sheet gently ease it to a standstill. Think of a car approaching a red light. The driver could simply tap the brakes gently for perhaps ten seconds, or jam them on hard at the last instant. Which do you think needs more energy?

• TAKE CARE! •

There's no need to remind anyone about the consequences of breaking an egg on a freshly cleaned sheet. You needn't really worry about that, but if you're feeling a little edgy you can practice a few times using a golf ball instead of an egg. Which comes with its own warning: don't do it anywhere near a window.

Foaming Finale

This experiment comes in at the very end for a good reason – it is the only one that also allows you and your crew to clear up afterwards. You could even say that the whole point of the experiment is to find a way of cleaning up after the experiment. Is that getting a little too confusing? Well, read on and prepare to do some scrubbing.

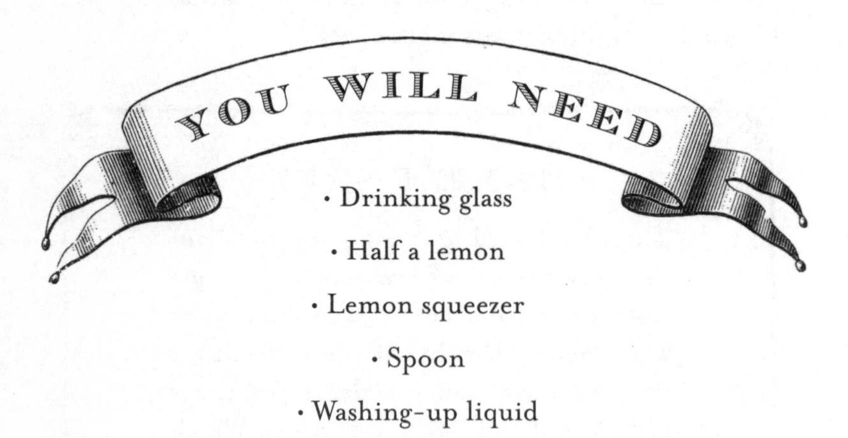

- Drinking glass
- Half a lemon
- Lemon squeezer
- Spoon
- Washing-up liquid
- 1 tsp bicarbonate of soda

METHOD

1. Add the bicarbonate of soda to the glass.

2. Squeeze in a good dollop (about 2 tsp) of washing-up liquid and stir.

3. Squeeze the juice from the lemon half.

4. Pour the lemon juice into the glass. Begin stirring.

5. Remove the spoon and observe the foam brewing up.

6. Stir some more to produce even more foam, which should start to ooze out over the edge of the glass.

7. Use the lemon-scented soapy foam to clean up your lab.

THE SCIENTIFIC EXCUSE

At its core, this experiment relies on an old standby – the reaction between the bicarbonate of soda (an alkali) and the citric acid of the lemon. When these combine, just like the combination of vinegar (another acid) and bicarbonate of soda, one of the by-products is carbon dioxide. And it's the CO_2 that keeps bubbling up.

TAKE CARE!

How can you possibly get into trouble when you've promised to clean up after yourself?

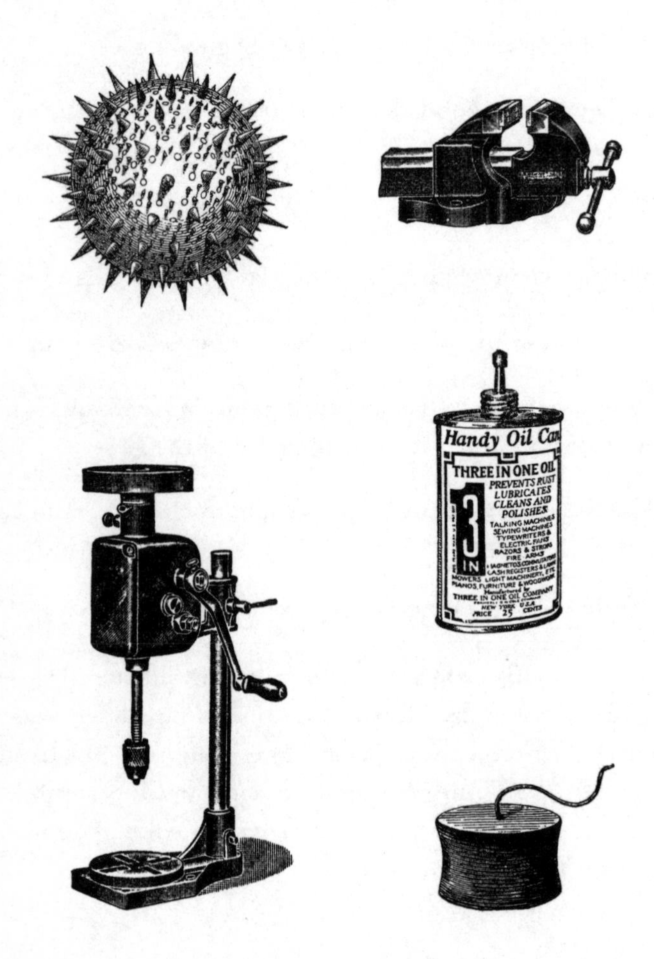
Handy Oil Can
THREE IN ONE OIL
3 IN 1
PREVENTS RUST
LUBRICATES
CLEANS AND
POLISHES
TALKING MACHINES
SEWING MACHINES
TYPEWRITERS &
ELECTRIC FANS
RAZORS & STROPS
FIRE ARMS
CASH REGISTERS & LAWN
MOWERS LIGHT MACHINERY, ETC
PIANOS, FURNITURE & WOODWORK
Manufactured by
THREE IN ONE OIL COMPANY
NEW YORK U.S.A.
PRICE 25 CENTS

At a glance

Some of the 65 experiments in this book can be done almost in an instant; others take up to several days to achieve their dramatic effect. The following list groups them in order of time taken, starting with those that can be done most quickly.

Flash in the Pan

less than 2 minutes

Five-minute wonders

2–5 minutes

On the hour

up to 1 hour

The 8-hour day

1–8 hours

Going the distance

a full day or more

ACKNOWLEDGEMENTS

I would like to thank the following individuals for their support and inspiration at every stage of this project. Their contributions – including direct participation, suggestions, reactions and scientific explanations – added depth and breadth to the book:

Janet Brakspear, Frank Ciccotti, Gregory Etter, Peter Filkins, Peter French, Dr Gary Hoffman, Benjamin Joyce, Dr Peter Lydon, William Matthiesen, Ian McChesney, Dr Sarah Morse, Peter Rielly, Susan Roeper, Elizabeth Stell.

In addition, I owe a debt of gratitude to the following companies and organisations:

Berkshire (Massachusetts) Film and Video, Camco International Ltd, The Corsham Bookshop, Energy for Sustainable Development Ltd, ESD Ventures, Kingswood School, the Met Office (UK), New York Public Library, Williams College.

Experiment Diary

Name of experiment:

Date:

Where conducted:

Weather conditions:

Witnesses:

Description of events and reaction:

Got into trouble?

Experiment Diary

Name of experiment:

Date:

Where conducted:

Weather conditions:

Witnesses:

Description of events and reaction:

Got into trouble?

Experiment Diary

Name of experiment:

Date:

Where conducted:

Weather conditions:

Witnesses:

Description of events and reaction:

Got into trouble?

Experiment Diary

Name of experiment:

Date:

Where conducted:

Weather conditions:

Witnesses:

Description of events and reaction:

Got into trouble?

Experiment Diary

Name of experiment:

Date:

Where conducted:

Weather conditions:

Witnesses:

Description of events and reaction:

Got into trouble?

Experiment Diary

Name of experiment:

Date:

Where conducted:

Weather conditions:

Witnesses:

Description of events and reaction:

Got into trouble?

Experiment Diary

Name of experiment:

Date:

Where conducted:

Weather conditions:

Witnesses:

Description of events and reaction:

Got into trouble?

Experiment Diary

Name of experiment:

Date:

Where conducted:

Weather conditions:

Witnesses:

Description of events and reaction:

Got into trouble?

Experiment Diary

Name of experiment:

Date:

Where conducted:

Weather conditions:

Witnesses:

Description of events and reaction:

Got into trouble?

Experiment League Table

(by category)

Shock value:

1.

2.

3.

4.

5.

Made people laugh:

1.

2.

3.

4.

5.

Easy to do:

1.

2.

3.

4.

5.

Best to do with friends:

1.

2.

3.

4.

5.

Learned something new:

1.

2.

3.

4.

5.

Would definitely do again:

1.

2.

3.

4.

5.

Experiment Checklist

Experiment	Ingredients found?	Venue chosen?
The flame tree?		
Going green – or is the green going?		
Pineapple power		
The peanut harvest		
Back-up plan		
The magic circle		
Cork on a bender		
The no-flow sieve		
Silent eruption		
The floating needle		
Ghostly skater		
The DIY ocean		
The paper polka		
Spooky calling card		
Crazy cereal		
Microwave soap		
Follow that star!		
Solar-powered oven		
Bending light		
The revenge of Icarus		
Impact craters		
Don't pressure me		
The sliding tumbler		
Bull-roarer		
Ping-pong pump		
The hankie dam		
Bottling out		
Air block!		
Weight of the world		
Balloon Hang Ten		
Heavy reading		
Straw rocket		
The can can-can		

Experiment Checklist

Date performed	Photos?	Experiment
		The flame tree?
		Going green – or is the green going?
		Pineapple power
		The peanut harvest
		Back-up plan
		The magic circle
		Cork on a bender
		The no-flow sieve
		Silent eruption
		The floating needle
		Ghostly skater
		The DIY ocean
		The paper polka
		Spooky calling card
		Crazy cereal
		Microwave soap
		Follow that star!
		Solar-powered oven
		Bending light
		The revenge of Icarus
		Impact craters
		Don't pressure me
		The sliding tumbler
		Bull-roarer
		Ping-pong pump
		The hankie dam
		Bottling out
		Air block!
		Weight of the world
		Balloon Hang Ten
		Heavy reading
		Straw rocket
		The can can-can

Experiment Checklist

Experiment	Ingredients found?	Venue chosen?
Create a cold front		
Being particular		
Storm in a … bottle		
Always take the weather		
Party pooper?		
The shy arsonist		
The tipsy ice cube		
That's tasteless!		
The folding egg		
Colour me confused		
The DNA detective		
Non-stick pan?		
Pepper pointers		
Glowing grapes		
Mad molecules		
Paper saucepan		
The disappearing beaker		
Dracula's cocktail		
The unbalanced balance		
The finger of fate		
Anti-flame thread		
Rule(r)s are made to be broken		
Skim reading?		
Zero-G cola		
Edible crystals		
Saloon brawl prop		
Cliffhanger ending		
Create a fossil		
Pharaoh's secret		
Cash and carry		
Scrambled bed		
Foaming finale		

Experiment Checklist

Date performed	Photos?	Experiment
		Create a cold front
		Being particular
		Storm in a ... bottle
		Always take the weather
		Party pooper?
		The shy arsonist
		The tipsy ice cube
		That's tasteless!
		The folding egg
		Colour me confused
		The DNA detective
		Non-stick pan?
		Pepper pointers
		Glowing grapes
		Mad molecules
		Paper saucepan
		The disappearing beaker
		Dracula's cocktail
		The unbalanced balance
		The finger of fate
		Anti-flame thread
		Rule(r)s are made to be broken
		Skim reading?
		Zero-G cola
		Edible crystals
		Saloon brawl prop
		Cliffhanger ending
		Create a fossil
		Pharaoh's secret
		Cash and carry
		Scrambled bed
		Foaming finale

WHOLLY
IRRESPONSIBLE
EXPERIMENTS!
PURE
NITRATE
OF
SODA
SEAN CONNOLLY

Enjoyed this book?
Want another chance to create some good honest mayhem on your journey of discovery?

Then look out for Sean Connolly's *Wholly Irresponsible Experiments!*, another 65 ways to get blown away by the wonders of science!

'This imaginative book is crammed with awe-inspiring experiments designed to demonstrate the magic of science.'
Guardian

'Hours of fun ... a marvellous way to excite children about the wonders of science'
Sunday Express

'In the gee-whiz spirit of *The Dangerous Book for Boys*'
Financial Times